THE OSKAR KLEIN MEMORIAL LECTURES

Volume 3

Editors

Lars Bergström & Ulf Lindström

Stockholm University

Sweden

World Scientific

New Jersey • London • Singapore • Hong Kong

Published by

World Scientific Publishing Co. Pte. Ltd.

P O Box 128, Farrer Road, Singapore 912805

USA office: Suite 1B, 1060 Main Street, River Edge, NJ 07661

UK office: 57 Shelton Street, Covent Garden, London WC2H 9HE

British Library Cataloguing-in-Publication Data
A catalogue record for this book is available from the British Library.

THE OSKAR KLEIN MEMORIAL LECTURES VOL. 3

ISBN 981-02-4691-9
ISBN 981-02-4692-7 (pbk)

Printed in Singapore.

Preface

The Oskar Klein Memorial Lecture series has been running for thirteen consecutive years since the start in 1988. In 1994, to commemorate the 100th anniversity of Klein's birth, the lectures were extended to a symposium.

The first four lectures, delivered by C.N. Yang, S. Weinberg, A. Bethe and A. Guth have been published in Volumes I and II of "The Oskar Klein Memorial Lectures", (ed. G. Ekspong) and the symposium is published as "The Oskar Klein Centenary" (ed. U. Lindström).

The present volume is thus the fourth in a series and has some expectations to live up to. Hopefully they are well met by the contributions by T.D. Lee, N. Seiberg, A. Polyakov, P.J.E. Peebles, E. Witten and G. 't Hooft.

The typical format of the Klein Lectures is that the speaker gives one general colloquium type lecture and one specialized seminar. We have included most of the colloquia and also some of the seminars. In the first two volumes it was still possible to also include some material from Klein's own research that had not been generally available. Here we have instead included more lectures. For this reason we have had to wait a while until there was enough material to fill a volume. However, the character of the lectures is such that they remain topical, and the outstanding quality of the contributors help to make these written versions timeless. In fact, one of the instructions we usually give to the speakers is to cover general aspects of their fields as well as the latest results. An example of this more timeless aspect is provided by the contribution by A. Polyakov where he has collected ideas on the parallels between different subjects that he has clearly pondered over some years.

The lecture series has contained some very exciting moments during the time covered in the present volume. We recall, e.g., the interesting discussion of symmetry by T.D. Lee, the presentation by N. Seiberg of wonderful new types of duality in supersymmetric theories, and by E. Witten of the then new ADS/CFT correspondence and its application to the long-standing problem of quark confinement, or the lecture in 1999 by G. 't Hooft, who went on to receive the Nobel Prize later on the same year.

Theoretical high-energy physics dominates the subjects of the lectures, mirroring one of Klein's own main interest. He had other interests as well, cosmology being one of them (particularly in his later work). It is therefore very befitting that the contibution by Peebles in a nice way summarizes the phenomenal improvement in the understanding of the Universe which has taken place since Klein's days. However, we also get sobering reminders that much still remains to be understood, and that it may be dangerous to overinterpret current results - a well-placed remark by one of the founders of physical cosmology.

All in all the yearly Klein Lecture has become a very sucessful tradition in Swedish physics and it has inspired several similar series. We are very grateful to the Swedish Royal Academy of Sciences whose support through its Nobel Institute for Physics has made these lectures possible over the years. We also take this opportunity to express our gratitude to FYSIKUM, Stockholm University, where the lectures and the traditional dinner are held.

Stockholm, April 2001

Lars Bergström and Ulf Lindström, editors.

CONTENTS

THE WEAK INTERACTION: ITS HISTORY AND IMPACT ON PHYSICS

T. D. Lee

Columbia University, New York, N.Y. 10027

It is a pleasure and an honor for me to give ths lecture in honor of Oscar Klein who made major contributions to field theory, quantum electrodynamcis and particle physics, including weak interactions. He was the first one to observe that the μ decay and the β decay could be described by the same interaction with the same coupling constant; this led to the discovery of the Universal Fermi Interaction.

Perhaps I should begin my discussion of the history of weak interactions by separating it into three periods:

1. Classical Period, 1898-1949
2. Transition Period, 1949-1956
3. Modern Period, 1956-

1. CLASSICAL PERIOD (β DECAY)

In 1898 Lord Rutherford[1] discovered that the so-called Becquerel ray actually consisted of two distinct components: one that is readily absorbed, which he called alpha radiation, and another of a more penetrating character, which he called beta radiation. With that began the history of the weak interaction. Then, in 1900[2], the Curies measured the electric charge of the β particle and found it to be negative.

Sometimes when we think of physics in those old days, we have the impression that life was more leisurely and physicists worked under less pressure. Actually, from the very start the road of discovery was tortuous and the competition intense. A letter written in 1902 by Rutherford (then 32) to his mother expressed the spirit of research at that time[3,4]:

"I have to keep going, as there are always people on my track. I have to publish my present work as rapidly as possible in order to keep in the race. The best sprinters in this road of investigation are Becquerel and the Curies... ."

Most of the people in this room can appreciate these words. Rutherford's predicament is still very much shared by us to this day. Soon many fast runners came: Hahn, Meitner, Wilson, Von Baeyer, Chadwick, Ellis, Bohr, Pauli, Fermi and many others.

In preparing this lecture, I was reminded once more of how relatively recent these early developments are. We know that to reach where we are today took more than 90 years and a large cast of illustrious physicists. I recall that when Lise Meitner came to New York in the mid '60s, I had lunch with her at a restaurant near Columbia. Later K.K. Darrow joined us. Meitner said, "It's wonderful to see young people." To appreciate this comment, you must realize that Darrow was one of the earliest members of the American Physical Society and at that lunch he was over 70. But Lise Meitner was near 90.

I was quite surprised when Meitner told me that she started her first postdoc job in theory with Ludwig Boltzmann. Now, Boltzmann was a contemporary of Maxwell. That shows us how recent even the "ancient" period of our profession is.

After Boltzmann's unfortunate death in 1906, Meitner had to find another job. She said she was grateful that Planck invited her to Berlin. However, upon arrival she found that because she was a woman she could only work at Planck's institute in the basement, and only through the servant's entrance. At that time, Otto Hahn had just set up his laboratory in an old carpenter shop nearby. Lise Meitner decided to join him and to become an experimentalist. For the next thirty years, their joint work shaped the course of modern physics.

In 1906, Hahn and Meitner published a paper[5] stating that the β ray carries a unique energy. Their evidence was that the absorption curve of a β ray shows an exponential decrease along its path when passing through matter, like the α ray.

Then W. Wilson[6], in 1909, said "no", the β ray does not have a unique energy. By observing the absorption curve through matter of an electron of unique energy, Wilson found electrons to exhibit totally different behavior from the α particle; the absorption curve of a unique energy electron is not exponential. Consequently, Wilson deduced that the apparent exponential behavior of the absorption curve of β decay implies that the β does not have a unique energy, the same experimental observation on β but with a totally opposite conclusion.

In 1910, Von Baeyer and Hahn[7] applied a magnetic field to the β ray; they found the β to have several discrete energies. In this way, they also reconciled the conclusion of W. Wilson. Then, in 1914, Chadwick[8] said "no". The β energy spans a continuous spectrum, instead of discrete values. The discrete energy observed by Von Baeyer and Hahn was due to the secondary electron from a nuclear γ transition, with the γ energy absorbed by the atomic electron. In this process, the discrete energy refers to the nuclear γ emission.

Then came World War I and scientific progress was arrested. In 1922, Lise Meitner[9] again argued that the β energy should be discrete, like α and γ. The apparent continuum manifestation is due to the subsequent electrostatic interaction between β and the nucleus. From 1922 to 1927, through a series of careful measurements, Ellis[10] again said "no" to Meitner's hypothesis. The β energy is indeed continuous. Furthermore, Ellis proved that the maximum β energy equals the difference of the initial and final nuclear energy.

There would then appear a missing energy. This was incorporated by Niels Bohr[11], who proposed the hypothesis of non-conservation of energy.

Very soon, Pauli said "no" to Bohr's proposition. Pauli[12] suggested that in the β decay energy is conserved, but accompanying the β particle there is always emission of a neutral particle of extremely small mass and with almost no interaction with matter. Since such a weakly interacting neutral particle is not detected, there appears to be an apparent nonconservation of energy.

Fermi[13] then followed with his celebrated theory of β decay. This in turn stimulated further investigation of the spectrum shape of the β decay, which did not agree with Fermi's theoretical prediction. This led Konopinski and Uhlenbeck[14] to introduce the derivative coupling. The confusion was only cleared up completely after World War II, in 1949, by Wu and Albert[15], signalling the end of one era and the beginning of a new one.

2. CLASSICAL PERIOD (OTHER WEAK INTERACTIONS)

When I began my graduate study of physics at the University of Chicago, in 1946, the pion was not known. Fermi and Teller[16] had just completed their theoretical analysis of the important experiment of Conversi, Pancini and Piccioni[17]. I attended a seminar by Fermi on this work. He cut right through the complex slowing-down process of the mesotron, the capture rate versus the decay rate, and arrived at the conclusion that the mesotron could not possibly be the carrier of strong forces hypothesized by Yukawa. Fermi's lectures were always superb, but that one to me, a young man not yet twenty and fresh from China, was absolutely electrifying. I left the lecture with the impression that, instead of Yukawa's idea, perhaps one should accept Heisenberg's suggestion[18] that the origin of strong forces could be due to higher-order processes of β interaction. As was known, these were highly singular.

At that time, the β interaction was thought to be reasonably well understood. Fermi's original vector-coupling form,

$$G\,(\psi_n^\dagger\,\gamma_4\gamma_\lambda\,\psi_p)\,(\psi_4^\dagger\,\gamma_4\gamma_\lambda\gamma_5\,\psi_\nu)$$

was, after all, too simple; to conform to reality, it should be extended to include a Gamow-Teller term. Fermi told me that his interaction was modelled after the electromagnetic forces between charged particles, and his coupling G was inspired by Newton's constant. His paper was, however, rejected by *Nature* for being unrealistic. It was published later in Italy, and then in *Zeitschrift für Physik*[13]. Fermi wrote his γ matrices explicitly in terms of their matrix elements. His lepton current differs from his hadron current by a γ_5 factor; of course the presence of this γ_5 factor has no physical significance. Nevertheless, it is curious why Fermi should choose this particular expression, which resembles the V-A interaction, but with parity conservation. Unfortunately, by 1956, when I noticed this, it was too late to ask Fermi.

A year later, the discovery of the pion through its decay sequence $\pi \to \mu \to e$ by Lattes, Muirhead, Occhialini and Powell[19] dramatically confirmed the original idea of Yukawa. The fact that the higher-order β interaction is singular is not a good argument that it should simply become the strong force.

This was then followed by Professor Klein's important discovery that μ decay and β decay can be described by the same four-fermion interaction. An excerpt from his writing is reproduced below.

No. 4101 June 5, 1948 NATURE 897

MESONS AND NUCLEONS

By Prof. O. KLEIN
Institut för Mekanik och Matematisk Fysik,
Stockholms Högskolas

Since, according to the above assumptions, the decay of the ordinary meson is, so to speak, the prototype of all β-processes, it is important that the value of the life-time, $\tau = 2 \times 10^{-6}$ sec., and the energy available in the process $\sim 100\ m_e c^2$, fit in very well with the value to be expected from our knowledge of the β-decay. Let ε be the energy available divided by $m_e c^2$, and $F(\varepsilon)$ the well-known Fermi function, which for large ε-values may be taken as $\varepsilon^5/50$. We then obtain for the product $\tau . F(\varepsilon)$— putting $\tau = 2 \times 10^{-6}$ sec. and $\varepsilon = 100$—the value 667, which is of about the expected magnitude.

. . . . . It was then found that the interaction constant corresponding to the meson life-time is of the same order of magnitude as that to be expected from Fermi's original theory of β-disintegration, a result which is, of course, identical with that of the above calculation

At Chicago we were not aware of Professor Klein's work. In January 1949 my fellow student, Jack Steinberger, submitted a paper[21] to *The Physical Review* in which he established that the μ meson disintegrates into three light particles, one electron and two neutrinos. This made it look very much like any other β decay, and stimulated Rosenbluth, Yang and myself to launch a systematic investigation. Are there other interactions, besides β decay, that could be described by Fermi's theory?

We found that if μ decay and μ capture were described by a four-fermion interaction similar to β decay, all their coupling constants appeared to be of the same magnitude. This was the beginning of the *Universal Fermi Interaction*. We then went on to speculate that, in analogy with electromagnetic forces, the basic weak interaction could be carried by a universal coupling through an *intermediate heavy boson*[22], which I later called $W^{\pm}$ for weak. Naturally I went to my thesis adviser, Enrico Fermi, and told him of our discoveries. Fermi was extremely encouraging. With his usual deep insight, he immediately recognized the further implications beyond our results. He put forward the problem that if this is to be the universal interaction, then there must be reasons why some pairs of fermions should have such interactions, and some pairs should not. For example, why does

$$p \not\rightarrow e^+ + \gamma$$

and

$$p \not\rightarrow e^+ + 2\nu\,?$$

A few days later, he told us that he had found the answer; he then proceeded to assign various sets of numbers, + 1, - 1 and 0, to each of these particles. This was the first time to my knowledge that both the laws of baryon-number conservation and of lepton-number conservation were formulated together to give selection rules. However, at that time (1948), my own reaction to such a scheme was to be quite unimpressed: surely, I thought, it is not necessary to explain why $p \not\rightarrow e + \gamma$, since everyone knows that the identity of a particle is never changed through the emission and absorption of a photon; as for the weak interaction, why should one bother to introduce a long list of mysterious numbers, when all one needs is to say that only three combinations $(\overline{n}p)$, $(\overline{e}\nu)$ and $\overline{\mu}\nu)$ can have interactions with the intermediate boson. (Little did I expect that soon there would be many other pairs joining these three.)

Most discoveries in physics are made because the time is ripe. If one person does not make it, then almost inevitably another person will do it at about the same time. In looking back, what we did in establishing the Universal Fermi Interaction was a discovery of exactly this nature. This is clear, since the same universal Fermi coupling observations were made independently by at least three other groups, Klein[20], Puppi[23], and Tiomno and Wheeler[24], all at about the same time. Yet Fermi's thinking was of a more profound nature. Unfortunately for physics, his proposal was never published. The full significance of these conservation laws was not realized until years later. While this might be the first time that I failed to recognize a great idea in physics when it was presented to me, unfortunately it did not turn out to be the last.

Thus, in 1949, there existed a simple theoretical frame based on the Fermi theory, describing the three weak interaction processes:

$$n \rightarrow p + e + \nu ,$$

$$\mu + p \rightarrow n + \nu$$

and

$$\mu \rightarrow e + 2\nu .$$

So, at the end of the classical period, we moved from the observation of β decay to the discovery of the Universal Fermi Interaction.

3. TRANSITION PERIOD (1949-1956)

Beginning in 1949, extensive work was done on the shape of the electron spectrum from μ decay. From the analysis of L. Michel[25], it was found that this distribution is given by

$$N(x) = x^2 \left\{ \left(2 - \frac{4}{3}\rho\right) - \left(2 - \frac{16}{9}\rho\right) x \right\}$$

where

$$x = (\text{momentum of } e)/(\text{maximum } e - \text{momentum}) ,$$

and ρ is the well-known Michel parameter, which can be any real number between 0 and 1, and measures the height of the end point at $x = 1$, as shown in Figure 1.

It is instructive to plot the experimental value of ρ against the year when the measurement was made. As shown in Figure 2, historically it began with $\rho = 0$ in 1949, at the beginning of the transition period. Then it slowly drifted upwards; only after the end of the transition period with the theoretical prediction in 1957 did it gradually become $\rho = \frac{3}{4}$. Yet, it is remarkable that at no time did the 'new' experimental value lie outside the error bars of the preceding one.

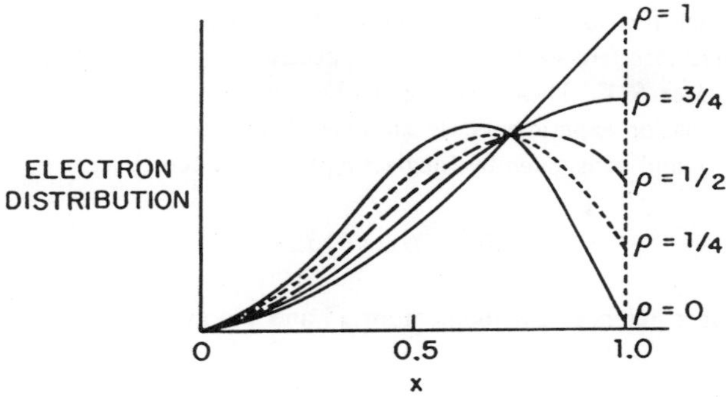

Figure 1. The ρ parameter in μ decay.

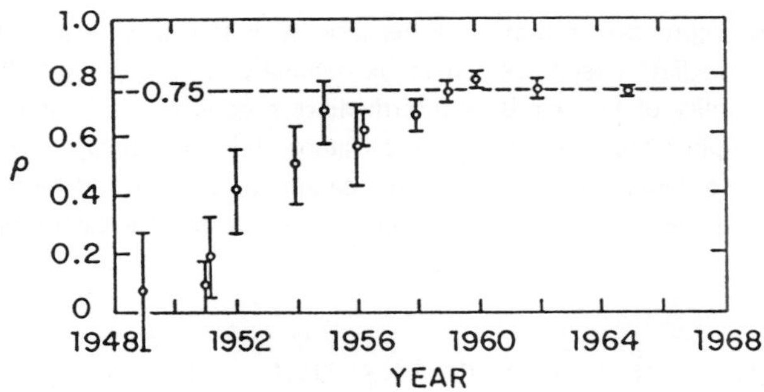

Figure 2. Variation of the ρ parameter over time.

In the same period (1949-56), a large amount of effort was also made on β decay experiments. By then, the Konopinski-Uhlenbeck interaction was definitely ruled out. The absence of the Fierz interference term[26] in the spectrum shows that the β interaction must be either V, A or S, T. These two possibilities were further resolved by a series of $\beta - \nu$ angular correlation experiments. In an allowed transition, the distribution for the angle θ between β and ν is given by (neglecting the Fierz term)

$$[1 + \lambda(P/E)_e \cos\theta]\, d\cos\theta\,,$$

where the subscript e refers to the momentum P and energy E of the electron. For a $\Delta J = 1$ transition,

$$\lambda = \begin{cases} +\frac{1}{3}, & \text{for } T \\ -\frac{1}{3}, & \text{for } A\,. \end{cases}$$

The experiment on ^{6}He decay by Rustad and Ruby gave[27]

$$\lambda = +0.34 \pm 0.09\,,$$

which seemed to establish unquestionably that the β decay interaction should be S, T with perhaps some unknown admixture of an additional pseudoscalar interaction.

I was quite depressed at that time because, with this new result, the theoretical idea of the intermediate boson seemed to be definitely ruled out. It is bad enough to assume the possiblity of two kinds of intermediate bosons of different spin-parity, one for the Fermi coupling and the other for the Gamow-Teller coupling. However, a tensor interaction with no derivative coupling simply cannot be transmitted by a spin-2 boson, since the former is described by an antisymmetric tensor and the latter by a symmetric one.

4. NEW HORIZON IN THE TRANSITION PERIOD

We now come to the $\theta - \tau$ puzzle.

During a recent physics graduate qualifying examination in a well-known American university, one of the questions was on the $\theta - \tau$ problem. Most of the students were puzzled over what θ was; of course they all knew that τ is the heavy lepton, the charged member of the third generation. So much for the history of physics.

In the early 1950s, θ referred to the meson which decays into 2π, whereas τ referred to the one decaying into 3π:

$$\theta \rightarrow 2\pi$$

and

$$\tau \rightarrow 3\pi\,.$$

The spin-parity of θ is clearly $0^+, 1^-, 2^+$, etc. As early as 1953, Dalitz[28] had already pointed out that the spin-parity of τ can be analyzed through his Dalitz plot and, by 1954, the then-existing data were more consistent with the assignment 0^- than 1^-. Although both mesons were known to have comparable masses (within ~ 20 MeV), there was, at that time, nothing too extraordinary about this situation. The masses of θ and τ are very near three times the pion mass, the phase space available for the θ decay is much bigger than that for τ decay; therefore one expects the θ decay rate to be much faster. However, when accurate lifetime measurements were made in 1955, it turned out that θ and τ have the same lifetime (within a few percent, which was the experimental accuracy). This, together with a statistically much more significant Dalitz plot of τ decay, presented a very puzzling picture indeed. The spin-parity of τ was determined to be 0^-; therefore it appeared to be definitely a different particle from θ. Yet, these two particles seemed to have the same lifetime, and also the same mass. This was the $\theta - \tau$ puzzle.

My first efforts were all on the wrong track. In the summer of 1955, Jay Orear and I proposed[29] a scheme to explain the $\theta - \tau$ puzzle within the bounds of conventional theory. We suggested a cascade mechanism, which turned out to be incorrect.

The idea that parity is perhaps not conserved in the decay of $\theta - \tau$ flickered through my mind. After all, strange particles are by definition strange, so why should they respect parity? The problem was that, after you say parity is not conserved in $\theta - \tau$ decay, then what do you do? Because if parity nonconservation exists only in $\theta - \tau$, *then we already have all the observable facts,* namely the same particle can decay into either 2π or 3π with different parity. I discussed this possibility with Yang, but we were not able to make any progress[30]. So we instead wrote papers on parity doublets, which was another wrong try[31].

5. THE BREAKTHROUGH (1956)

The Rochester meeting on high energy physics was held from April 3 to 7, 1956.

At that time, Steinberger and others were conducting extensive experiments on the production and decay of the hyperons Λ^0 and Σ^-:

$$\pi^- + p \rightarrow \begin{cases} \Lambda^0 + K^0 \\ \Sigma^- + K^+ \end{cases} \tag{1}$$

and

$$\left\{ \begin{matrix} \Lambda^0 \\ \Sigma^- \end{matrix} \right\} \rightarrow \pi + N\,. \tag{2}$$

The dihedral angle ϕ between the production plane and the decay plane (which will be defined below) is of importance for the determination of the hyperon spin.

Let $\vec{\pi}$, $\vec{\Lambda}$ and $\vec{N}$ be the momenta of π, Λ in process (1) and N in (2), all, say, in the respective center-of-mass systems of the reactions. The normal to the production plane is parallel to $\vec{\pi} \times \vec{\Lambda}$, and that to the decay plane to $\vec{\Lambda} \times \vec{N}$. Hence the dihedral angle ϕ is defined through its cosine:

$$\cos\phi \ \propto \ (\vec{\pi} \times \vec{\Lambda}) \cdot (\vec{\Lambda} \times \vec{N})\,. \tag{3}$$

Its distribution is

$$D(\phi) \ = \ \begin{cases} 1 & \text{if the hyperon-spin is } \frac{1}{2}\,, \\ 1 + \alpha \cos^2\phi & \text{if the hyperon-spin is } \frac{3}{2}\,, \end{cases} \tag{4}$$

etc. By this definition, ϕ varies from 0 to π. Furthermore, $D(\phi)$ is identical to $D(\pi - \phi)$. At the Rochester Conference, Jack Steinberger gave a talk and plotted his data on $D(\phi)$ with ϕ varying from 0 to π. However other physicists, W.D. Walker and R.P. Shutt, plotted $D(\phi) + D(\pi - \phi)$; in this way ϕ can only vary from 0 to $\frac{\pi}{2}$. After the conference, Jack came to my office to discuss a letter which he had just received from R. Karplus. In this letter Karplus questioned why Jack did not join the others, since the total number of events was (at that time) quite limited, and a folding of $D(\pi - \phi)$ onto $D(\phi)$ would increase the experimental sensitivity of the spin determination. Jack wanted to know how certain was the relationship that $D(\phi)$ is an even function of $\cos\phi$.

The dihedral angle, as defined by expression (3), has nothing to do with parity, since it is a scalar. In the course of explaining to Jack the $\cos^2\phi$ dependence of $D(\phi)$, I suddenly realized that if one changes the definition of ϕ to be the angle of rotation around the Λ momentum-vector, which is the intersection of these two planes, then the range of ϕ can be extended from 0 to 2π; that is, in place of (3), one defines ϕ through the pseudoscalar

$$\sin\phi \ \propto \ (\vec{\pi}_\perp \times \vec{N}_\perp) \cdot \vec{\Lambda}, \tag{5}$$

where $\vec{\pi}_\perp$ and $\vec{N}_\perp$ refer to the components of $\vec{\pi}$ and $\vec{N}$ perpendicular to $\vec{\Lambda}$, as shown in Figure 3. In this case, ϕ can vary from 0 to 2π.

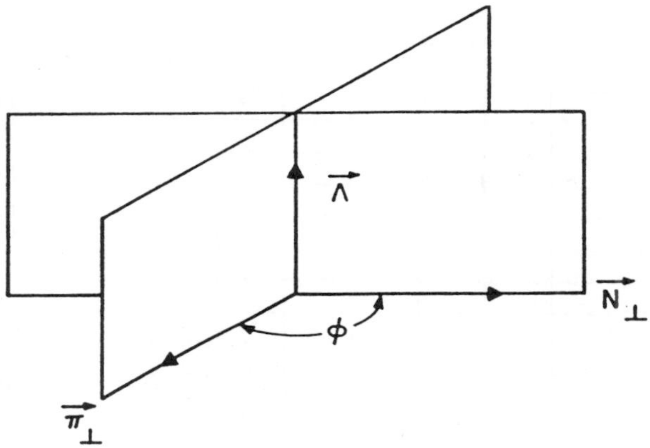

Figure 3. The dihedral angle ϕ between the production plane and the decay plane.

If parity is not conserved in strange particle decays, there could be an asymmetry between events with ϕ from 0 to π and those with ϕ from π to 2π. *This is the missing key!* I was quite excited, and urged Jack to re-analyse his data immediately and test the idea experimentally. This led to the very first experiment on parity nonconservation. Very soon, within a week, Jack and his collaborators (Budde, Chrétien, Leitner, Samios and Schwartz) had their results, and the data were published[32] even before the theoretical paper[33] on parity nonconservation. The odds turned out to be 13 to 3 in Σ^- decay and 7 to 15 in Λ^0 decay (see Figure 4).

Of course, because of the limited statistics, no definitive conclusion on parity violation could be drawn. Nevertheless, had the statistics been ten times more, then with the same kind of ratio one could have made a decisive statement on parity conservation. This showed clearly that parity violation could be tested experimentally provided one measured a pseudoscalar, such as (5).

However, on the theoretical side there was still the question of parity conservation in ordinary β decay. In this connection, at the beginning of May, C.N. Yang came to see me and wished to join me in the examination of β decay. This led to our discovery that, in spite of the extensive use of parity in nuclear physics and β decay, there existed no evidence at all of parity conservation in any weak interaction.

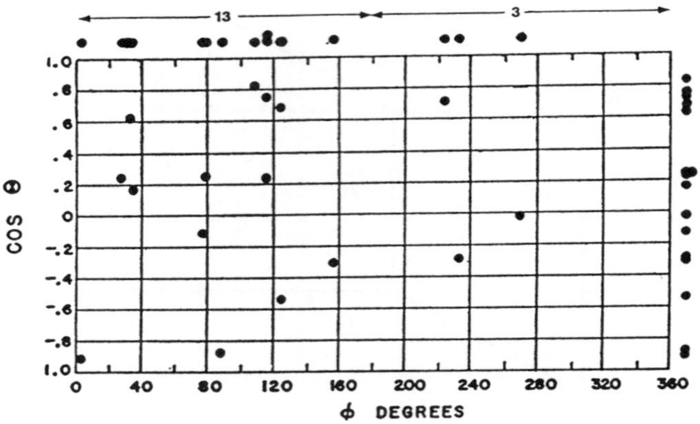

Figure 4a. **Angular correlation plot of the cosine of the polar angle against the azimuthal angle ϕ for the reaction $\pi^- + p \rightarrow \Sigma^- + K^+$, $\Sigma^- \rightarrow \pi^- + n$. The point histograms on the edges of the figure represent the data after integration over the other coordinate. (Taken from Ref. 32)**

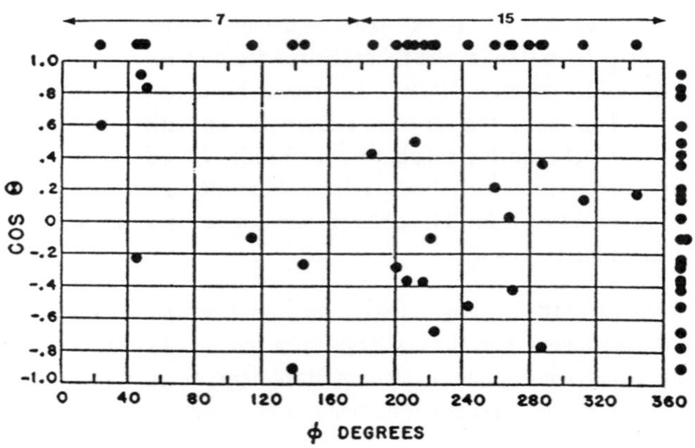

Figure 4b. **Angular correlation plot for the reaction $\pi^- + p \rightarrow \Lambda^0 + \theta^0$, $\Lambda^0 \rightarrow \pi^- + p$. (Taken from Ref. 32)**

T.D.L. and C.N. Yang,
"Question of Parity Nonconservation in Weak Interactions,"
Phys.Rev. **104**, 254 (1956).
"A Two-component Theory of the Neutrino,"
Phys.Rev. **105**, 1671 (1957).
T.D.L., R. Oehme and C.N. Yang,
"Possible Noninvariance of T, C and CP,"
Phys.Rev. **106**, 340 (1957).
C.S. Wu, E. Ambler, R.W. Hayward, D. Hoppes and R.P. Hudson,
"Experimental Test of Parity Nonconservation in Beta Decay,"
Phys.Rev. **105**, 1413 (1957).
R.L. Garwin, L.M. Lederman and M. Weinrich,
"Nonconservation of P and C in Meson Decays,"
Phys.Rev. **105**,1415 (1957).
J.I. Friedman and V.L. Telegdi,
"Parity Nonconservation in $\pi^+ - \mu^+ - e^+$,"
Phys.Rev. **105**,1681 (1957).
M. Goldhaber, L. Grodzins and A.W. Sunyar,
"Helicity of Neutrinos,"
Phys.Rev. **109**, 1015 (1958).
R.E. Marshak and E.C.G. Sudarshan,
"Chirality and the Universal Fermi Interaction,"
Phys.Rev. **109**, 1860 (1958).
R. Feynman and M. Gell-Mann,
"Theory of the Fermi Interaction,"
Phys.Rev. **109**, 193 (1958).
R. Christenson, J. Cronin, V.L. Fitch and R. Turlay,
"2π Decay of the K_2° Meson,"
Phys.Rev.Lett. **13**, 138 (1964).

Table 1.

Several months later followed the decisive experiments by Wu, Ambler, Hayward, Hoppes and Hudson, at the end of 1956, on β decay[34], and by Garwin, Lederman and Weinrich[35] and by Friedman and Telegdi[36] on $\pi - \mu$ decay. Table 1 lists the important papers on symmetry violation in weak interactions.

6. SYMMETRY VIOLATIONS

Wu *et al.* investigated the decay of polarized cobalt nuclei, Co^{60}, into electrons. Because these nuclei are polarized they rotate parallel to each other. The experiment consisted of two setups, identical except that the directions of rotation of the initial nuclei were opposite; that is, each was a looking-glass image of the other. The experimenters found, however, that the patterns of the final electron distributions in these two setups are not mirror images of each other. In short, the initial states are mirror images, but the final configurations are not (see Figure 5). This established parity nonconservation, i.e., right-left asymmetry.

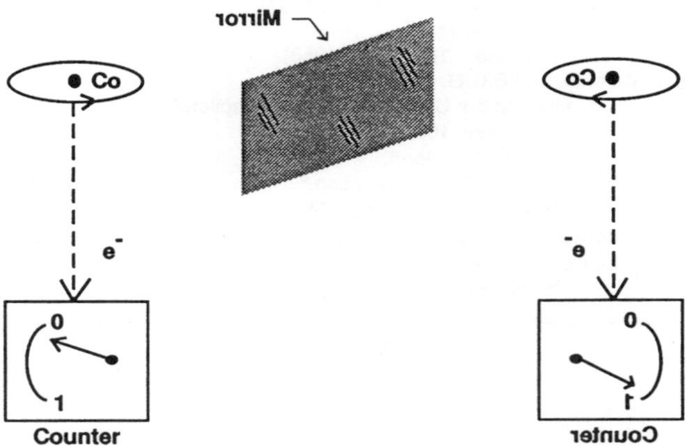

Figure 5. The initial setups of these two experiments on Co^{60} decay are exact mirror images,but the final electron distributions are not, as indicated by the different readings on the counters.

The β decay experiment by Wu *et al.* and the $\pi - \mu$ decay experiment by Garwin, Lederman and Weinrich and by Friedman and Telegdi proved not only parity P violation, but also asymmetry under particle-antiparticle conjugation C. This is illustrated in Figure 6. Both the neutrino and the antineutrino possess a spin (angular momentum). For a neutrino, if one aligns one's left thumb parallel to its momentum, then the curling of one's four fingers would always be in the direction of its spin, indicating P violation. Therefore, the spin and momentum direction of a neutrino defines a perfect left-hand screw, whereas the spin and momentum direction of an antineutrino defines a perfect right-hand screw, showing C violation in addition to P violation. This property holds for neutrinos and antineutrinos everywhere, independently of how they are produced.

In 1964, the experiment by Christenson *et al.*[37] established CP violation. An example of how such a violation can be established experimentally is to examine the decay[38] of K_L^0 into $e^+\pi^-\nu$ and $e^-\pi^+\bar{\nu}$. The long-lived neutral kaon K_L^0 is a spherically symmetric particle with zero spin; it carries no electric charge and no electromagnetic form factor of any kind. Yet these two decay modes have different rates:

$$\frac{\text{rate}(K_L^0 \to e^+ + \pi^- + \nu)}{\text{rate}(K_L^0 \to e^- + \pi^+\bar{\nu})} = \mathbf{1.00648} \pm 0.00035\,.$$

Consequently, by using a time-clocking device it is possible to differentiate the positive sign of electricity vs. the negative sign of electricity, showing C violation. Because this rate difference remains true under mirror reflection, CP is also violated. The present status of these discrete symmetries and asymmetries is given in Table 2.

At present, only the combined symmetries CPT remain valid. In other words, only when we interchange

$$\text{particle} \leftrightarrow \text{antiparticle},$$

$$\text{right} \leftrightarrow \text{left},$$

$$\text{past} \leftrightarrow \text{future},$$

do all physical laws appear to be invariant. From CPT symmetry, it follows that CP violation also implies time reversal T asymmetry.

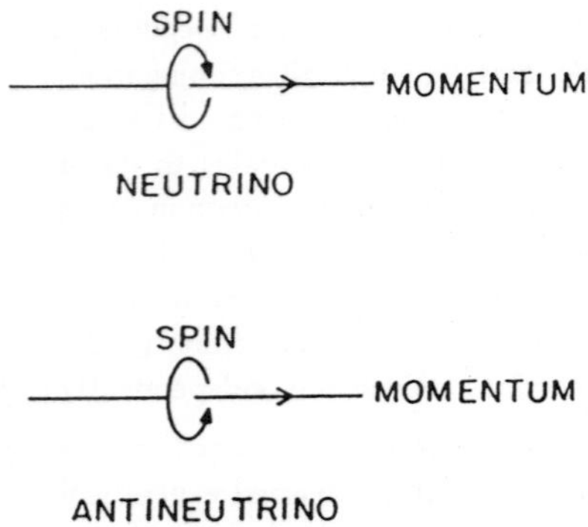

Figure 6. The spin momentum of a neutrino defines a left-hand screw; that of an antineutrino defines a right-hand screw.

Present Status	
CPT	Good
P	X
C	X
CP	X
T	X ← without assuming CPT
CT	X
PT	X

Table 2.

7. TIME REVERSAL

Time reversal symmetry T means that the time-reversed sequence of any motion is also a possible motion. Some of you may think this absurd, since we are all getting older, never younger. So why should we even contemplete that the laws of nature should be time-reversal symmetrical?

In this sense we must distinguish between the evolution of a small system and a large system. Let me give an example. In Figure 7a each circle represents an airport, and a line indicates an air corridor. We assume that between any two of these airports the number of flights going both ways along any route is the same (this property will be referred to as microscopic reversability). Thus a person in "X" can travel to Stockholm (for this discussion we assume the only air connection from "X" is to Stockholm), then through Stockholm to Paris or New York. At any point in his travel, he can return to "X" with the same ease. But suppose that in every airport we were to remove all the signs and flight information, while maintaining exactly the same number of flights, as shown in Figure 7b. A person starting from "X" would still arrive in Stockholm, since that is the only airport connected to "X". However, without the signs to guide him, it would be very difficult for him to pick out the return flight to "X" from the many gates in the Stockholm airport. The plane he gets on may be headed for New York. If, in New York, he then tries another plane again without any guidance, he could perhaps arrive in Tokyo. If he keeps on going this way, his chance of getting back to "X" is very slim indeed. In this example, we see that microscopic reversibility is strictly maintained. When all the airport destination signs and other flight information are given clearly, then macroscopically we also have reversibility. On the other hand, if all such information is withheld, then the whole macroscopic process appears irreversible. Thus, macroscopic irreversibility is not in conflict with microscopic reversibility.

Time reversal symmetry in physics refers to *microscopic reversiblity* between all molecular, atomic, nuclear and subnuclear reactions. Since none of these molecules, atoms, nuclei or subnuclear particles can be easily marked, any microscopic system in nature would exhibit irreversibility. This result is independent of microscopic reversbility. In any macroscopic process, we have to average over an immense number of unmarked microscopic units of atoms, molecules, and so forth (as in the example of unmarked airports and air routes), and that gives rise to *macroscopic irreversibility.* It is in this statistical sense of ever increasing disorder (entropy) that we define the direction of our macroscopic time flow. We may recall the words from *H.M.S. Pinafore*, by Gilbert and Sullivan,

What never? No, never!
What never? Well, hardly ever!

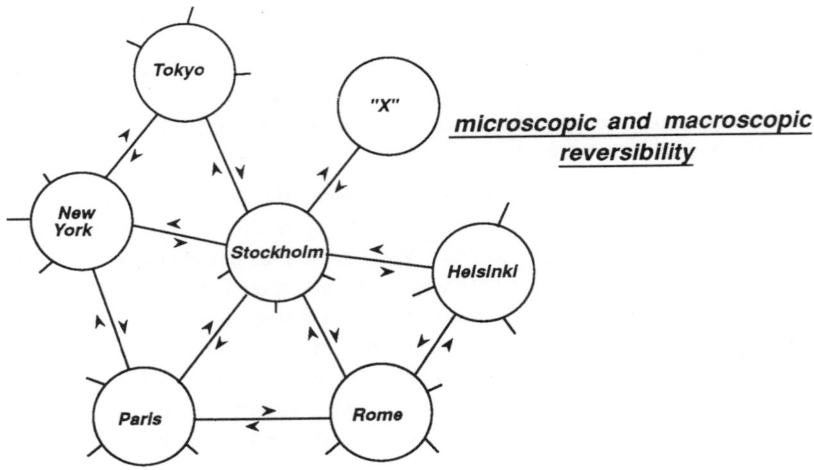

Figure 7a. In this example, microscopic reversibility means an equal number of flights in either direction on every air route. When the names of the airports, the numbers of the gates, and all flight information are known, there is also macroscopic reversibility.

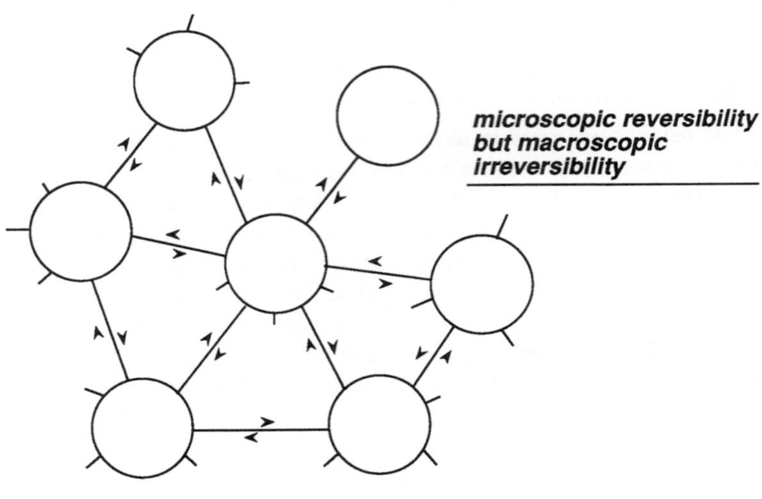

Figure 7b. If we maintain the same number of flights in each direction on any route (i.e., microscopic reversibility), but remove destination signs, gate numbers, and all other information, then it is nearly impossible to find our way back (i.e., macroscopic irreversibility).

The existence of the macroscopic time direction then leaves open the important question whether time-reversal symmetry, or microscopic reversibility, is true or not. Since 1964, after the discovery of CP violation by Christenson *et al.*, through a series of remarkable experiments involving the kaons, it was found that the microscopic reversibility is indeed violated. Nature does not seem to respect time-reversal symmetry!

8. PRESENT STATUS

In 1967, the paper of S. Weinberg[39] unified the weak interaction with the electromagnetic force, called the electroweak interaction, which is mediated by the photon γ, the charged intermediate bosons $W^{\pm}$ and their neutral partner Z^0 (these particles were discovered at CERN in 1983 [40]).

At present, our theoretical structure of electroweak, strong and gravitational forces can be summarized as follows:

QCD (strong interaction)

$SU(2) \times U(1)$ **Theory** (electro-weak)

General Relativity (gravitation).

However, in order to apply these theories to the real world, we need a set of about 17 parameters, all of unknown origins. Thus, this theoretical edifice cannot be considered complete.

The two outstanding puzzles that confront us today are:

i) Missing symmetries - All present theories are based on symmetry, but most symmetry quantum numbers are *not* conserved.

ii) Unseen quarks - All hadrons are made of quarks; yet, no individual quark can be seen.

As we shall see, the resolutions of these two puzzles probably are tied to the structure of our physical vacuum.

The puzzle of missing symmetries implies the existence of an entirely new class of fundamental forces, the one that is responsible for symmetry breaking. Of this new force, we know only of its existence, and very little else. Since the masses of all known particles break these symmetries, an understanding of the symmetry-breaking forces will lead to a comprehension of the origin of the masses of all known particles. One of the promising directions is the spontaneous symmetry-breaking mechanism in which one assumes that the physical laws remain symmetric, but the physical vacuum is not. If so, then the solution of this puzzle is closely connected to the structure of the physical vacuum; the excitations of the physical vacuum may lead to the discovery of Higgs-type mesons.

In some textbooks, the second puzzle is often "explained" by using the analogy of the magnet. A magnet has two poles, north and south. Yet, if one breaks a bar magnet in two, each half becomes a complete magnet with two poles. By splitting a magnet open one will never find a single pole (magnetic monopole). However, in our usual description, a magnetic monopole can be considered as either a fictitious object (and therefore unseeable) or a real object but with exceedingly heavy mass beyond our present energy range (and therefore not yet seen). In the case of quarks, we believe them to be real physical objects and of relatively low masses (except the top quark); furthermore, their interaction becomes extremely weak at high energy. If so, why don't we ever see free quarks? This is, then, the real puzzle.

The fact that quarks and gluons cannot be seen individually suggests an entirely new direction in our understanding of particles. Traditionally, any particle is either stable or unstable; the former is represented by a pole on the physical sheet and the latter by a pole on the "second" sheet. But now, we have a new third class: particles that cannot be seen individually. This is illustrated in Table 3, and its solution in Table 4. Following the KLN Theorem,[41] gluons and quarks are indeed discovered by observing high energy jets. The second method, of using entropy change, will be discussed shortly. Before doing that, we may recall what has happened in the past.

At the end of the last century, there were also two physics puzzles:

1. No absolute inertial frame (Michelson-Morley Experiment 1887),
2. Wave-particle duality (Planck's formula 1900).

These two seemingly esoteric problems struck classical physics at its very foundation. The first became the basis for Einstein's theory of relativity and the second laid the foundation for us to construct quantum mechanics. In this century, all the modern scientific and technological developments—*nuclear energy, atomic physics, molecular structure, lasers, x-ray technology, semiconductors, superconductors, supercomputers*—only exist because we have relativity and quantum mechanics. To humanity and to our understanding of nature, these are all-encompassing.

Now, near the end of the twentieth century, we must ask what will be the legacy we give to the next generation in the next century? At present, like the physicists at the end of the 1890's, we are also faced with two profound puzzles. It seems likely that the present two puzzles may bring us as important a change in the development of science and technology in the twenty-first century.

STABLE PARTICLES POLES ON THE "PHYSICAL" SHEET	UNSTABLE PARTICLES POLES ON THE "SECOND" SHEET	PARTICLES THAT CANNOT BE SEEN INDIVIDUALLY
GRAVITON γ ν e p :	μ π K : n J/ψ Υ : W Z :	GLUONS QUARKS :

Table 3.

HOW TO DETECT PARTICLES THAT CANNOT BE SEEN INDIVIDUALLY?
1. JETS (KLN THEOREM ON MASS SINGULARITIES (1964)) 2. ENTROPY DENSITY CHANGE (VACUUM EXCITATIONS T.D.L. AND G.C.WICK (1974))

Table 4.

9. PHYSICAL VACUUM

The current explanation of the quark confinement puzzle is again to invoke the vacuum. We assume the QCD vacuum to be a condensate of gluon pairs and quark-antiquark pairs so that it is a perfect color dia-electric[42] (i.e., color dielectric constant $\kappa = 0$). This is in analogy to the description of a superconductor as a condensate of electron pairs in BCS theory, which results in making the superconductor a perfect dia-magnet (with magnetic susceptibility $\mu = 0$). When we switch from QED to QCD we replace the magnetic field $\vec{H}$ by the color electric field $\vec{E}_{\text{color}}$, the superconductor by the QCD vacuum, and the QED vacuum by the interior of the hadron. Just as the magnetic field is expelled outward from the superconductor, the color electric field is pushed into the hadron by the QCD vacuum, and that leads to color confinement, or the formation of bags.[43] This situation is summarized below.

QED superconductivity as a perfect dia-magnet		QCD vacuum as a perfect color dia-electric
$\vec{H}$	$\longleftrightarrow$	$\vec{E}_{\text{color}}$
$\mu_{\text{inside}} = 0$	$\longleftrightarrow$	$\kappa_{\text{vacuum}} = 0$
$\mu_{\text{vacuum}} = 1$	$\longleftrightarrow$	$\kappa_{\text{inside}} = 1$
inside	$\longleftrightarrow$	outside
outside	$\longleftrightarrow$	inside

In the resolution of both puzzles, missing symmetry and quark confinement, the system of elementary particles no longer forms a self-contained unit. The microscopic particle physics depends on the coherent properties of the macroscopic world, represented by the appropriate operator averages in the physical vacuum state.

If we pause and think about it, this represents a rather startling conclusion, contrary to the traditional view of particle physics which holds that the microscopic world can be regarded as an isolated system. To a very good approximation it is separate and uninfluenced by the macroscopic world at large. Now, however, we need these vacuum averages; they are due to some long-range ordering in the state vector. At present our theoretical technique for handling such coherent effects is far from being developed. Each of these vacuum averages appears as an independent parameter, and that accounts for the large number of constants needed in the present theoretical formulation.

On the experimental side, there has hardly been any direct investigation of these coherent phenomena. This is because hitherto in most high-energy experiments, the higher the energy the smaller has been the spatial region we are able to examine. In order to explore physics in this fundamental area, relativistic heavy ion collisions offer an important new direction.[44] The basic idea is to collide heavy ions, say gold on gold, at an ultra-relativistic region. Before the collision, the vacuum between the ions is the usual physical vacuum; at a sufficiently high energy, after the collision almost all of the baryon numbers are in the forward and backward regions (in the center-of-mass system). The central region is essentially free of baryons and, for a short duration, it is of a much higher energy density than the physical vacuum. Therefore, the central region represents the excited vacuum (Figure 8).

As we shall see, we need RHIC, the 100 GeV $\times$ 100 GeV (per nucleon) relativistic heavy ion collider at the Brookhaven National Laboratory, to explore the QCD vacuum.

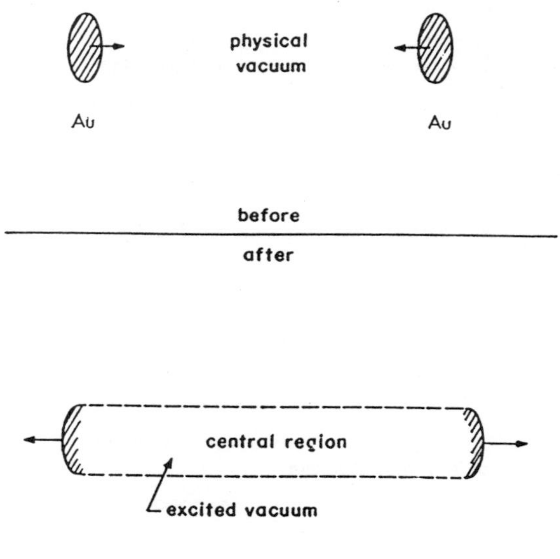

Figure 8. Vacuum excitation through relativistic heavy ion collisions.

10. PHASE DIAGRAM OF THE QCD VACUUM

A normal nucleus of baryon number A has an average radius $r_A \approx 1.2A^{\frac{1}{3}}fm$ and an average energy density

$$\mathcal{E}_A \approx \frac{m_A}{(4\pi/3)r_A^3} \approx 130MeV/fm^3 .$$

Each of the A nucleons inside the nucleus can be viewed as a smaller bag which contains three relativistic quarks inside; the nucleon radius is $r_N \approx 0.8fm$ and its average energy density is

$$\mathcal{E}_N \approx \frac{m_N}{(4\pi/3)r_N^3} \approx 440MeV/fm^3 .$$

Consequently, even without any sophisticated theoretical analysis we expect the QCD phase diagram to be of the form given by Figure 9.

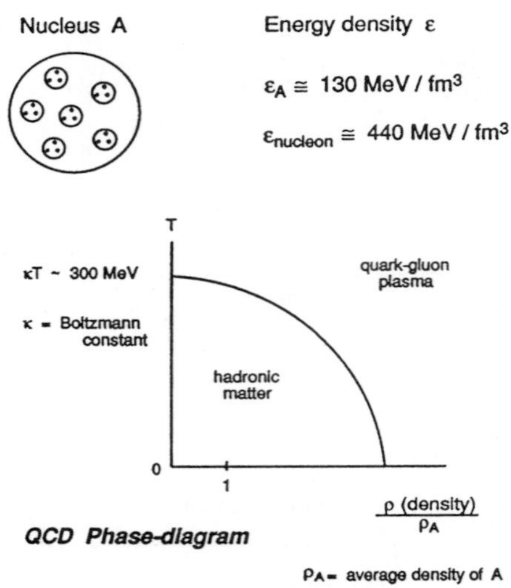

Figure 9.

In Figure 9, the ordinate is κT ($\kappa =$ Boltzmann constant, $T =$ temperature), the abscissa is ρ/ρ_A ($\rho =$ nucleon density, $\rho_A =$ average nucleon density in a normal nucleus A). A typical nucleus A is when $\rho = \rho_A$. The scale can be estimated by noting that the critical $\kappa T \sim 300$ MeV is about the difference of $1 fm^3$ times $\mathcal{E}_N - \mathcal{E}_A$ and the critical $\rho/\rho_A \sim 4$ is just the nearest integer larger than $(1.2/0.8)^3$.

Accurate theoretical calculation exists only for pure lattice QCD (i.e., without dynamical quarks). The result is shown in Figure 10.

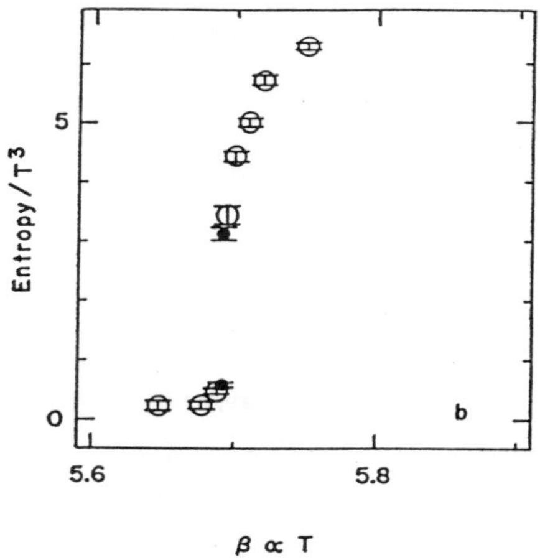

Figure 10. Phase transition[45] (pure QCD)

If one assumes scaling, then the phase transition in pure QCD (zero baryon number, $\rho = 0$) occurs at $\kappa T \sim 340$ MeV with the energy density of the gluon plasma

$$\mathcal{E}_P \sim 3GeV/fm^3 .$$

To explore this phase transition in a relativistic heavy ion collision, we must examine the central region. Since only a small fraction of the total energy is retained in the central region, it is necessary to have a beam energy (per nucleon) at least an order of magnitude larger than $\mathcal{E}_N \times (1.2fm)^3 \sim 5$ GeV; this makes it necessary to have an ion collider of 100 GeV $\times$ 100 GeV (per nucleon) for the study of the QCD vacuum.

Another reason is that at 100 GeV $\times$ 100 GeV the heavy nuclei are almost transparent, leaving the central region (in Figure 10) to be one almost without any baryon number. As remarked before, this makes it an ideal situation for the study of the excited vacuum. Suppose that the central region does become a quark-gluon plasma. How can we detect it? This will be discussed in the following.

11. $\pi\pi$-INTERFEROMETRY

As shown in Figure 11, the emission amplitude of two pions of the same charge with momenta $\vec{k}_1$ and $\vec{k}_2$ from points $\vec{r}_1$ and $\vec{r}_2$ is proportional to

$$A \equiv e^{i\vec{k}_1\cdot\vec{r}_1+i\vec{k}_2\cdot\vec{r}_2} + e^{i\vec{k}_2\cdot\vec{r}_1+i\vec{k}_1\cdot\vec{r}_2},$$

because of Bose statistics. Let

$$\vec{q} \equiv \vec{k}_1 - \vec{k}_2 \qquad \text{and} \qquad \vec{r} \equiv \vec{r}_1 - \vec{r}_2 .$$

Hanbury - Brown / Twiss - type Experiment

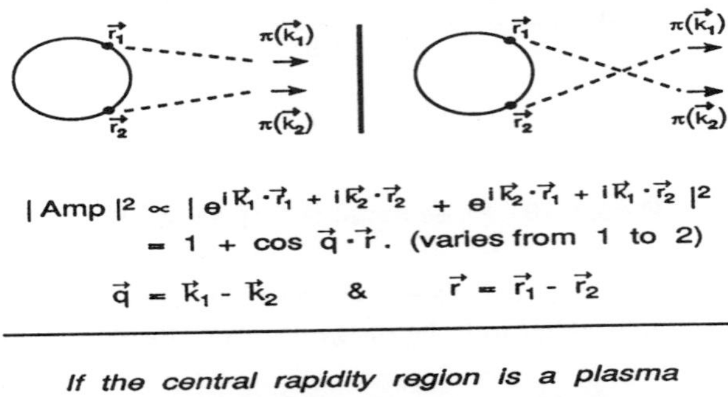

$$|\,\text{Amp}\,|^2 \propto |\, e^{i\vec{k}_1\cdot\vec{r}_1 + i\vec{k}_2\cdot\vec{r}_2} + e^{i\vec{k}_2\cdot\vec{r}_1 + i\vec{k}_1\cdot\vec{r}_2}\,|^2$$
$$= 1 + \cos \vec{q}\cdot\vec{r}. \text{ (varies from 1 to 2)}$$
$$\vec{q} = \vec{k}_1 - \vec{k}_2 \qquad \& \qquad \vec{r} = \vec{r}_1 - \vec{r}_2$$

If the central rapidity region is a plasma of entropy density S_{plasma}, which later hadronizes (of entropy density S_{Had}),

$$(\text{vol} \cdot S)_{plasma} \leq (\text{vol} \cdot S)_{Had}$$
$$S_{Had} << S_{plasma} \quad \therefore \quad (\text{vol})_{Had} >> (\text{vol})_{plasma}$$

Figure 11

Since

$$|A|^2 \equiv 1 + \cos \vec{q} \cdot \vec{r}$$

changes from $|A|^2 = 2$ as $\vec{q} \to 0$, to $|A|^2 = 1$ as $\vec{q} = \infty$, a measurement of the $\pi\pi$ correlation gives a determination of the geometrical size R of the region that emits these pions, like the Hanbury-Brown/Twiss determination of the stellar radius.

Now, if the central region is a plasma of entropy density S_P occupying a volume V_P, which later hadronizes to ordinary hadronic matter (of entropy density S_H and volume V_H), the total final entropy $S_H V_H$ must be larger than the total initial entropy $S_P V_P$. Since $S_P > S_H$, we have

$$V_H > V_P .$$

The experimental configurations and results[46] are given in Figure 12 and Table 5. One sees that the hadronization radius in the central region is indeed much larger than that in the fragmentation region. This is, at best, only indicative of the quark-gluon plasma. Much work and higher energy are needed for a more definitive proof. Nevertheless, it does show that relativistic heavy ion can be an effective means of exploring the structure of the vacuum.

12. CONCLUDING REMARKS

To conclude we emphasize, once again, that the most challenging problems in physics are

(1) the symmetry-breaking force, and

(2) the structure of the vacuum.

It is quite likely that the answer to these two problems lies in the same direction. They can only be solved when we learn how to excite the vacuum. In the traditional way of thinking, our world is the world of particles. Larger units are made of small ones, which in turn are made of even smaller elements. The search for the smallest building block that everything is made of drives us to explore physical phenomena within smaller and smaller distances; that necessitates energies higher and higher in inverse proportion to the distance in question. On the other hand, the puzzles of missing symmetry and quark confinement have forced us to face the profound possibility that the vacuum could be a physical medium.

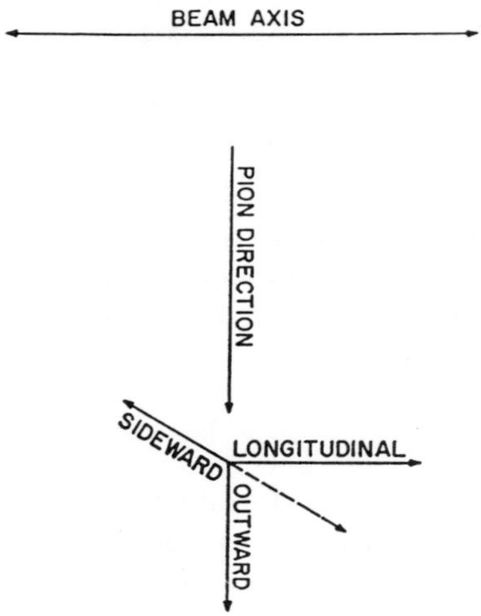

Figure 12

Rapidity Interval	Gaussian R_T(fm)	R_L (fm)	Λ
$1 < y < 2$	4.3 ± 0.6	2.6 ± 0.6	$0.34^{+0.09}_{-0.06}$
	$R_T^{side} = 4.0 \pm 1.0$ fm	2.6 ± 0.6	$0.34^{+0.09}_{-0.06}$
	$R_T^{out} = 4.4 \pm 1.0$ fm	2.6 ± 0.6	$0.34^{+0.09}_{-0.06}$
$2 < Y < 3$ (Central region (mid-rapidity))	8.1 ± 1.6	$5.6\ ^{+1.2}_{-0.8}$	0.77 ± 0.19
	$R_T^{side} = 6.6 \pm 1.8$ fm	$5.6\ ^{+1.2}_{-0.8}$	0.77 ± 0.19
	$R_T^{out} = 11.2 \pm 2.3$ fm	$5.6\ ^{+1.2}_{-0.8}$	0.77 ± 0.19

R (Oxygen) $\cong$ 3 fm

Table 5. $\pi\pi$-interference result[46] from the collision of an O beam (200 GeV/nucleon) on a stationary Au target

As we look into the future, the completion of RHIC in 1997 offers an unprecedented opportunity for physicists to explore the possibility of exciting the vacuum and to examine whether it is indeed a physical medium.

If the vacuum is the underlying cause for the strange phenomena in the microscopic world of particle physics, it must also have been actively responsive to the macroscopic distribution of matter and energy in the universe. Because the vacuum is everywhere and forever, these two, the micro- and the macro-, have to be linked together; neither can be considered a separate entity. Future history books will record that ours was a time when humankind was able to forge this bond on a scientific basis.

This research was supported in part by the U.S. Department of Energy

REFERENCES

[1] E. Rutherford, Philos.Mag. **42**, 392 (1898).

[2] M. and P. Curie, C.R. Acad.Sci. **130**, 647 (1900).

[3] A.S. Eve, *Rutherford* (Cambridge, Cambridge University Press, 1939), p. 80.

[4] A. Pais, *Inward Bound* (Oxford, The Clarendon Press, 1985).

[5] O. Hahn and L. Meitner, Phys.Z. **9**, 321, 697 (1908).

[6] W. Wilson, Proc.Roy.Soc. **A82**, 612 (1909).

[7] O. von Baeyer and O. Hahn, Phys.Z. **11**, 488 (1910).

[8] J. Chadwick, Verh.Dtsch.Phys.Ges. **16**, 383 (1914).

[9] L. Meitner, Z.Phys. **9**, 131, 145 (1922).

[10] E.D. Ellis, Proc.Cambridge Philos.Soc. **21**, 121 (1922).

[11] N.Bohr, J.Chem.Soc. **135**, 349 (1932).

[12] W. Pauli, American Physical Society Meeting in Pasadena, June 1931.

[13] E. Fermi, Ric.Scient. **4**, 491 (1934); Nuovo Cimento **11**, 1 (1934); Z.Phys. **88**, 161 (1934).

[14] E.J. Konopinski and G.E. Uhlenbeck, Phys.Rev. **48**, 7 (1935).

[15] C.S. Wu and R.D. Albert, Phys.Rev. **75**, 315 (1949).

[16] E. Fermi and E. Teller, Phys.Rev. **72**, 399 (1947).

[17] M. Conversi, E. Pancini and O. Piccioni, Phys.Rev. **68**, 232 (1945).

[18] W. Heisenberg, Z.Phys. **101**, 533 (1936).

[19] C.M.G. Lattes, H. Muirhead, G.P.S. Occhialini and C.F. Powell, Nature **159**, 694 (1947).

[20] O. Klein, Nature **161**, 897 (1948).

[21] J. Steinberger, Phys.Rev. **75**, 1136 (1949).

[22] T.D. Lee, M. Rosenbluth and C.N. Yang, Phys.Rev. **75**, 905 (1949).

[23] G. Puppi, Nuovo Cimento **6**, 194 (1949).

[24] J. Tiomno and J.A. Wheeler, Rev.Mod.Phys. **21**, 153 (1949).

[25] L. Michel, Nuovo Cimento **10**, 319 (1953).

[26] M. Fierz, Z.Phys. **104**, 553 (1937).

[27] B.M. Rustad and S.L. Ruby, Phys.Rev. **89**, 880 (1953).

[28] R.H. Dalitz, Philos.Mag. **44**, 1068 (1953); Phys.Rev. **94**, 1046 (1954).

[29] T.D. Lee and J. Orear, Phys.Rev. **100**, 932 (1955).

[30] *High Energy Nuclear Physics*, Proc. 6th Annual Rochester Conference, April 3-7 1956 (New York, Interscience, 1956), p. VI-20.

[31] T.D. Lee and C.N. Yang, Phys.Rev. **102**, 290 (1956).

[32] R. Budde, M. Chrétien, J. Leitner, N.P. Samios, M. Schwartz and J. Steinberger, Phys.Rev. **103**, 1827 (1956).

[33] T.D. Lee and C.N. Yang, Phys.Rev. **104**, 254 (1956).

[34] C.S. Wu, E. Ambler, R.W. Hayward, D.D. Hoppes and R.P. Hudson, Phys.Rev. **105**, 1413 (1957).

[35] R.L. Garwin, L.M. Lederman and M. Weinrich, Phys.Rev. **105**, 1415 (1957).

[36] J.I. Friedman and V.L. Telegdi, Phys.Rev. **105**, 1681 (1957).

[37] J.H. Christenson, J.W. Cronin, V.L. Fitch and R. Turlay, Phys.Rev.Lett. **13**, 138 (1964).

[38] C. Alff-Steinberger *et al.*, Phys.Lett. **20**, 207 (1966). See also *Particles and*

Detectors, Festschrift for Jack Steinberger, eds. K. Kleinknecht and T.D. Lee (Berlin, Springer Verlag, 1986), p. 285ff for other references.

[39] S. Weinberg, Phys.Rev.Lett. **19**, 1264 (1967). See also S.L. Glashow, Nucl.Phys. **22**, 579 (1961); A. Salam and J.C. Ward, Nuovo Cimento **11**, 568 (1959), Phys.Lett. **13**, 168 (1964); A. Salam, in *Proceedings of the 8th Nobel Symposium*, ed. N. Svartholm (Stockholm, Almqvist and Wiksell, 1968), p. 367.

[40] C. Rubbia, in *Proceedings of the International Conference on High Energy Physics*, eds. J. Guy and C. Costain (Didcot, Rutherford Appleton Laboratory, 1983), p. 880.

[41] T. Kinoshita, J.Math.Phys. **3**, 650 (1950). T.D. Lee and M. Nauenberg, Phys.Rev. **133**, B1549 (1964).

[42] R. Friedberg and T.D. Lee, Phys.Rev. D**18**, 2623 (1978).

[43] A. Chodos, R.J. Jaffe, K. Johnson, C.B. Thorn and V.F. Weisskopf, Phys.Rev. D**9**, 3471 (1974).

[44] T.D. Lee and G.C. Wick, Phys.Rev. D**9**, 2291 (1974); T.D. Lee, "Relativistic Heavy Ion Collisions and Future Physics," in *Symmetries in Particle Physics*, ed. I. Bars, A. Chodos and C.H. Tze (New York, Plenum Press, 1984), p. 93.

[45] F.R. Brown, N.H. Christ, Y.F. Deng, M.S. Gao and T.J. Woch, Phys.Rev.Lett. **61**, 2058 (1988).

[46] W. Willis, "Experimental Studies on States of the Vacuum," in *Relativistic Heavy-Ion Collisions*, ed. R.C. Hwa, C.S. Gao and M.H. Ye (New York, Gordon and Breach Science Publishers, 1990), p. 39.

ELECTRON ORBITS AND SUPERCONDUCTIVITY OF CARBON 60

T. D. Lee

Columbia University, New York, N.Y. 10027

1. Introduction

The discovery of C_{60} and the superconductivity of its alloy compounds[1–4] provides a rich field for new theoretical and experimental investigations.[5–8] In this paper, we show that the near degeneracy of the low-lying vector (T_{1u}) and axial-vector (T_{1g}) levels of an isolated C_{60}^- may provide a simple pairing mechanism in C_{60}^{--}, C_{60}^{---} and the K_nC_{60} crystal.

We will begin with a review of the single-electron energy levels of a C_{60} molecule. Above the 60-closed shell lie two triplets possessing the same icosahedral symmetry label T_1 but of opposite parity. We shall derive their wave functions in the tight-binding limit of the molecular orbit approximation, and exhibit these functions in a simple and useful form which prompts us to refer to them as "vector" and "axial vector". Our analysis is then extended to C_{60}^{--}. By taking into account the strong Coulomb energy between electrons, we find that the spectrum of C_{60}^{--} contains two low-lying states invariant under the proper icosahedral symmetry transformations, but again of opposite parity and hence naturally called "scalar" and "pseudoscalar", with an energy degeneracy within 1% of their excitation energy (in our approximate calculation), even closer than the vector axial-vector doublets in C_{60}^-.

Since parity doublets, two states of opposite parity with nearly the same energy, are not a common occurrence in physics, it is worthwhile to examine the mechanism that gives rise to them in C_{60}. For a microscopic system, such as mesons and baryons, composite particles of opposite parities usually have quite different internal structures which make them unlikely to be degenerate. In most macroscopic systems, as in the case of left-handed or right-handed sugar molecules, it is conceptually simple to contstruct coherent mixtures of opposite parities that are nearly degenerate; the difficulty lies in the practical realization of such coherent mixtures.

2. Icosahedral Group

In order to have the necessary tools for our subsequent analysis, we give a brief discussion of the molecular orbits of C_{60}. As will be shown below, these can be expressed in terms of the regular representation of the proper

$$\text{icosahedral group } \mathcal{G} = \{g\} \tag{1}$$

where g denotes its sixty group elements. Corresponding to the identity element e, we may take any point $\hat{e}$ on a unit sphere with center 0. The application of the finite rotation that associates with g transforms $\hat{e}$ to $\hat{g}$. For $\hat{e}$ not lying on any of the symmetry axes, the resulting sixty $\hat{g}$ are all different. (A convenient notation is to use the same $\hat{e}$ and $\hat{g}$ to denote the corresponding unit radial vectors of the sphere, as well as their end points on the sphere.) For definiteness, let the unit sphere be the *circumsphere* of a regular icosahedron of vertices $N, A_1, \cdot\cdot, A_5', S$. As in Figure 1, choose the point e (corresponding to the identity element e in $\mathcal{G}$) on the edge $\overline{A_1N}$, with $\overline{eN} = \frac{1}{3}\,\overline{A_1N}$ and $\overrightarrow{0e}\|\hat{e}$. Each unit radial vector $\hat{g}$ intersects a point g on one of the edges of the icosahedron. The location of each point g, thus generated, gives the position of a carbon nucleus in C_{60}, provided the unit scale of the radius is $R \cong 4.03\mathring{A}$, so that the circumspherical radius of C_{60} is its actual value $R_0 \cong 3.5\mathring{A}$, with the ratio

$$\frac{R}{R_0} = 3\left(5+\frac{4}{\sqrt{5}}\right)^{-\frac{1}{2}} \cong 1.15\,. \tag{2}$$

(The same notation g denotes both the group element as well as the position of the carbon nucleus.) Any function $f(g)$ of these sixty positions g can be expressed in terms of the irreducible representations of the icosahedral group $\mathcal{G}$, which consist of a singlet **1**, two triplets **3** and $\tilde{\mathbf{3}}$, a quartet **4** and a quintet **5**. This gives

$$60 = 1^2+3^2+3^2+4^2+5^2\,. \tag{3}$$

In the literature[9,10], these representations are often referred to as A, T_1, T_2, G and H respectively. Our notations follow those used in particle physics, with the dimensionality of the irreducible representation shown explicitly.

The irreducible representations **1**, **3** and **5** can be readily identified with the usual $\ell = 0$ (s-wave), $\ell = 1$ (p-wave) and $\ell = 2$ (d-wave) of the spherical harmonics $Y_{\ell,m}(\theta,\phi)$, where $m = -\ell,\ -\ell+1, \cdot\cdot, \ell$, and θ, ϕ are the polar and azimuthal angles. To derive the remaining two irreducible representations $\tilde{\mathbf{3}}$ and **4**, we designate one of the icosahedral vertices as the north pole N ($\theta = 0$), as in Figure 1. Express any function F of θ, ϕ defined on the twelve vertices of the icosahedron in terms of the sixteen spherical

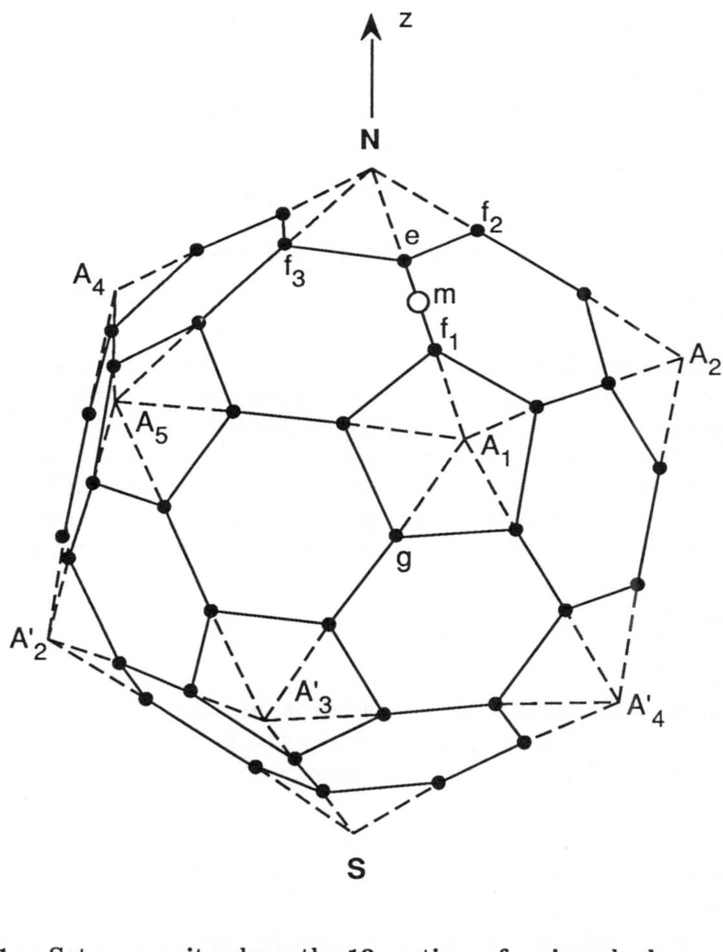

Figure 1. **Set on a unit sphere the 12 vertices of an icosahedron:** N, S, A_i **and** A_i' **(i = 1, 2, ·, 5) with** $\overline{NS}$, $\overline{A_iA_i'}$ **the diameters of the sphere whose center is 0. Each of the sixty dots** $e, f_1, f_2, f_3, g, \cdots$ **denote the positions of C atoms in** C_{60}**. The three nearest neighbors of** e **are** f_1, f_2 **and** f_3**. The distance** $\overline{ef_2} = \overline{ef_3}$ **is slightly different from** $\overline{ef_1}$**.**

harmonics: $\ell = 0, 1, 2$ and 3 of $Y_{\ell,m}(\theta,\phi)$. (Note that $16 = 1 + 3 + 5 + 7$.) On the other hand, the function F has only

$$12 = 1+3+5+3 \tag{4}$$

values. Hence, we expect four linear combinations of the seven $\ell = 3$ $Y_{\ell,m}(\theta,\phi)$ to be identically zero on these twelve icosahedral vertices, leaving a triplet $\tilde{\mathbf{3}}$. For an explicit construction, we observe that the geodesic arc between two nearest neighboring vertices of an icosahedron is $\cos^{-1}(1/\sqrt{5})$; therefore,

$$Y_{3,\pm1}(\theta,\phi) \propto (5\cos^2\theta - 1)\sin\theta\, e^{\pm i\phi} \tag{5}$$

is zero on all these twelve vertices. Likewise, on these sites

$$Y_{3,\pm3}(\theta,\phi) \propto Y_{3,\mp2}(\theta,\phi), \tag{6}$$

from which we can readily form two linear combinations of these $\ell = 3$ spherical harmonics that are also zero. In (6), both sides have the same azimuthal variation because of the five-fold symmetry along the north pole-south pole diameter; hence, (6) holds in the northern hemisphere, since excluding the pole, all the other five icosahedral vertices are at the same latitude. Its validity in the southern hemisphere then follows from inversion symmetry. In this way, we see that under the icosahedral rotations, the seven $Y_{3,m}(\theta,\phi)$ functions decompose into a quartet $\mathbf{4}$ and another triplet $\tilde{\mathbf{3}}$, different from $\mathbf{3}$.

3. C_{60}

Returning to the C_{60} structure on the unit sphere, we denote by $|g>$ a vector (*ket*-vector in Dirac's notation) with sixty components. Each element g corresponds to a definite finite rotation of the icosahedral group.

The radial position vector $\hat{g}$ of each C atom on the sphere corresponds to an element g of the proper icosahedral group. Thus the 60-dimensional Hilbert space of the tight-binding limit supports the regular representation of this group. For a typical "hopping" Hamiltonian of icosahedral symmetry, the lowest single-particle state is that having constant wave function, reminiscent of the s-wave (i.e., the orbital angular momentum quantum number $\ell = 0$) on a sphere. In the icosahedral group this singlet representation $\mathbf{1}^+$ is usually labeled A, or more fully A_g (the subscript $g =$ *gerade* meaning even parity).

The next-lowest state has a threefold degeneracy (excluding spin) which may be described as follows. For each C atom it turns out[11] that there is a certain unit vector

$\hat{e}_0(g)$, very close to $\hat{g}$, within an angle of about 1°; the wave function at site $\hat{g}$ can be chosen to be proportional to $\hat{e}_0(g) \cdot \hat{n}$ where $\hat{n}$ is some fixed vector. Clearly one obtains in this way three independent wave functions of the same energy, corresponding to the three independent choices of $\hat{n}$. If we approximate $\hat{e}_0(g) \cong \hat{g}$ (as we shall do hereafter in this discussion), then these three states correspond to the p-wave triplet ($\ell = 1$) on the sphere. Under the icosahedral group this triplet representation $\mathbf{3}^-$ is usually labeled T_1, or more fully T_{1u} ($u = ungerade$ meaning odd parity, since $\hat{g}$ changes sign for oppositely placed C atoms).

It is natural to call the singlet and triplet states described above "scalar" and "vector" referring to the way they transform when the laboratory coordinate system is rotated or inverted. Thus, the designation "vector" refers to the vector $\hat{n}$ which selects a T_1 state. Since multiplicity and parity play a prominent role in our analysis, we shall write $\mathbf{1}^+$(scalar) for A_g, $\mathbf{3}^-$(vector) for T_{1u}, etc., with $\mathbf{1}^+$ denoting a singlet of even parity, $\mathbf{3}^-$ a triplet of odd parity, etc. (All these multiplicities are doubled by spin.)

In the regular representation, each irreducible representation of dimension d appears d times; thus, the triplet T_1 must appear three times. Our interest will be in the *other* two triplets T_1, whose energy lies far above that of the "p-wave" state described above. Their wave functions can be expressed simply as follows:

Because of the detailed structure of the molecule, each C atom defines not only a radial vector $\hat{g}$ but a local coordinate system, with its three orthonormal basis vectors denoted by

$$\hat{g}, \qquad \hat{e}_-(g) \qquad \text{and} \qquad \hat{e}_+(g)\,, \tag{7}$$

where

$$\hat{e}_+(g) \;=\; \hat{g} \times \hat{e}_-(g)\,. \tag{8}$$

Under an inversion, $\hat{g}$ changes sign; consequently, $\hat{e}_-(g)$ and $\hat{e}_+(g)$ are of opposite parities. Let $\hat{e}_-(g)$ be of odd parity, then $\hat{e}_+(g)$ is of even parity. By taking the wave function proportional to $\hat{e}_-(g) \cdot \hat{n}$ (where, as before, $\hat{n}$ is some fixed vector) one obtains a triplet $\mathbf{3}^-$ or T_{1u}, the LUMO (lowest unoccupied molecular orbital) state in neutral C_{60}. By taking $\hat{e}_+(g) \cdot \hat{n}$, one obtains a $\mathbf{3}^+$ or T_{1g}, the second LUMO state in neutral C_{60}. It is natural to call these triplets "vector" and "axial-vector" respectively. Together, (7) determines the set of three triplets T_1 contained in the regular representation.

To transform a C atom to one of its nearest neighbors by an element of the icosahedral group, one must rotate through a large angle (180° for one of the nearest neighbors, from e to f_1 in Figure 1), about an axis rather close to $\hat{g}$. Hence $\hat{g}$ varies slowly between neighbors but, as we shall see, $\hat{e}_-(g)$ and $\hat{e}_+(g)$ vary rapidly and, because of (8), by about the same amount. This is why the $\hat{e}_- \cdot \hat{n}$ and $\hat{e}_+ \cdot \hat{n}$ triplets have energy far above that of the $\hat{g}\cdot\hat{n}$ triplet, but close to each other. (Although the high-lying $\mathbf{3}^-$ and $\mathbf{3}^+$ states have the same formal structure under the *icosahedral group* as the low-lying $\mathbf{3}^-$, if one attempts to fill in a smooth wave function between the C atoms, one will need mostly spherical harmonics of $\ell = 5$ and $\ell = 6$ for $\hat{e}_-$ and $\hat{e}_+$ respectively, instead of $\ell = 1$ as for $\hat{g}$.)

Since $\hat{e}_- \cdot \hat{n}$ and $\hat{e}_+ \cdot \hat{n}$ are the two low-lying levels above the 60-closed shell, in C_{60}^- the extra electron may be in either of these triplets. The nearness of these two levels of opposite parity, combined with their compatibility (both T_1) under the icosahedral group, render C_{60}^- highly polarizable; the polarizability can be readily calculated.

4. Pairing Mechanism

In C_{60}^{--}, the Coulomb energy between the two extra electrons plays a dominant role. Without this term, the lowest-energy state would be of the form $T_{1u}T_{1u}$ or $\mathbf{3}^- \times \mathbf{3}^-$. But we find that two other states have considerably lower Coulomb energy. One is obtained by combining T_{1u} and T_{1g} to make a two-particle state $\mathbf{1}^-$:

$T_{1u}T_{1g} + T_{1g}T_{1u}$ with wave function

$$[\hat{e}_-(g)\cdot\hat{e}_+(g') + \hat{e}_+(g)\cdot\hat{e}_-(g')](\uparrow\downarrow' - \downarrow\uparrow'). \tag{9}$$

The other is made by mixing the $\mathbf{1}^+$ combination of $T_{1u}T_{1u}$ with that of $T_{1g}T_{1g}$:

$T_{1u}T_{1u} + T_{1g}T_{1g}$ with wave function

$$[\hat{e}_-(g)\cdot\hat{e}_-(g') - \hat{e}_+(g)\cdot\hat{e}_+(g')](\uparrow\downarrow' - \downarrow\uparrow'), \tag{10}$$

where $\hat{g}$, $\uparrow$ or $\downarrow$ and $\hat{g}'$, $\uparrow'$ or $\downarrow'$ are the position and spin variables of the two electrons. Because both wave functions vanish when the two electrons coalesce, $\hat{g} = \hat{g}'$, their mutual Coulomb energy is greatly reduced; this leads to the spin-0 parity doublets in C_{60}^{--}. labeled $\mathbf{1}^-$ and $\mathbf{1}^+$.

The wave functions (9) and (10) represent singlets in both position and spin; no external vector $\hat{n}$ appears. It is natural to call these paired states pseudoscalar and scalar. It is evident that if the energy difference $\Delta\epsilon$ between $\mathbf{3}^-$ (T_{1u}) and $\mathbf{3}^+$ (T_{1g}) is small, then it will contribute only to order $(\Delta\epsilon)^2$ to the splitting between scalar and pseudoscalar paired states. This is why the paired states in C_{60}^{--} form a much tighter parity doublet than the component one-particle states in C_{60}^-.

The closeness of these parity doublets makes it natural to isolate their response to strong interactions, separate from other distant levels. This provides a convenient means to derive analytical expressions for many of the important parts of strong interaction effects. Thus, we can calculate the polarization energy of C_{60}^{-} in a strong electric field E, showing that it changes from the weak field expression $-\frac{1}{2}\alpha E^2$ to one that depends linearly on E. Another example is to derive the final energy in C_{60}^{--}, within the parity doublet approximation, to all powers of the interaction Hamiltonian.

5. K_3C_{60}

We turn our attention to the K_3C_{60} crystal and calculate the Madelung energy. The result shows that all K are ionized, as is commonly accepted.

Above the 60-closed shell, a typical band calculation reveals a low-lying cluster of three overlapping narrow bands, which is the Bloch wave extension of the three components of the vector wave function $\hat{e}_-(g)$ of an isolated C_{60}^{-}. These overlapping fermion bands are half-filled in the case of K_3C_{60}. On the other hand, the pseudoscalar pairing wave function (9) in C_{60}^{--} suggests a different Bloch wave extension, one that represents the hopping of such a highly correlated two-particle state in the crystal. The orthogonality of these correlated two-particle Bloch wave functions to any product of two one-particle wave functions in the fermion bands follows from the original orthogonality condition in a single C_{60} molecule:

$$\sum_{g=1}^{60} \hat{e}_-(g)_i \, \hat{e}_+(g)_j = 0 \tag{11}$$

where i and j denote the vector components. We call the Bloch extension of (9) the "boson band" (or the pseudoscalar band).

The two electrons in the fermion bands have an energy advantage over the boson on account of the excitation energy $\Delta\epsilon$ and the lowering in kinetic energy; however, there is a disadvantage to the fermion bands of having a higher Coulomb energy. The important question concerning the role of bosons versus fermions depends on the delicate balance between these two opposing factors. This is examined in Ref. 11. We start with the Coulomb energy ($\sim$ 11eV per lattice cell) between three electrons in the fermion bands, and compare that with the configuration of placing two electrons in the boson band and the remaining one in the fermion bands. We then take into account the kinetic energy difference and the Van der Waals energy difference; the latter is important because of the large polarizability of the C_{60} negative ion. The final energy balance between these two configurations is estimated to be only 0.22eV, slightly in favor of the fermion bands. However, considering the highly approximate nature of our calculations and that the final answer is only about 2% of the initial energy that we begin with, it is not possible to make any definitive statement. Nevertheless, a variety of interesting theoretical possibilities emerges. It seems likely that there exists a narrow boson band, which may lie very close to, or overlapping with, the three fermion bands.

The possibility of a close-by low-lying boson band, in addition to the usual fermion bands, has important consequences for superconductivity. The bosons may undergo Bose-Einstein condensation. The zero momentum nature of the Bose condensate necessarily generates charge fluctuations in the coordinate space; thereby it increases the Coulomb energy. It can be shown that this Coulomb energy increase is compensated for, at near distances, by the monopole-dipole interaction between neighboring C_{60} molecules because of their large polarizability. At large distances there is, in addition, the Debye screening of the Coulomb potential generated by these charge fluctuations.

In Ref.11, through a simplified but explicit field theoretic model, we are able to examine the effect of the Bloch extension of the correlated scalar wave function (10), the parity doublet partner of the pseudoscalar (9). Except in the somewhat unlikely case that the boson band is lower in energy than two times the bottom energy of the fermion bands, the scalar only provides a resonance to the electrons in the fermion bands. Such a resonance can produce an energy gap, as in the BCS theory of superconductivity. While both members of the bosonic parity doublet can be important to superconductivity, their roles are different. The experimentally observed pressure variation of the critical temperature[6] can be shown to be consistent with the model.

Our thesis is that the $\mathbf{3}^+$ (T_{1g}) level plays a special role because of its coupling and near degeneracy with the $\mathbf{3}^-$ (T_{1u}) at the Fermi level. Therefore the C_{60}^{--} energy should be calculated by starting with the mixed states of (9) and (10), and treating all levels *outside* the parity doublet perturbatively. This makes it possible to handle strong interactions, and leads to high T_c superconductivity.

The parity doublets provide an essential insight into the nature of the paired wave-function in C_{60}, as well as the physics associated with polarizability of the C_{60} negative ion. The superconductivity is discussed on the basis of a simplified model Hamiltonian in which there is an effective "attractive and local" four-fermion field interaction term. Its presence is due to the lowering of the strong Coulomb energy in the correlated paired states (9) and (10) in C_{60}^{--} which, in turn, can be justified within the parity doublet approximation. On the other hand, parity doublets are rather special to C_{60}; a natural question is to ask how important are their roles to superconductivity in general. We believe that, for all high T_c superconductors (cupric oxides and C_{60} alloys) with very small coherence length, the essential common feature probably lies in the approximate applicability of an effective "attractive and local" interaction term. For material other than C_{60}, its underlying reason may have nothing to do with parity doublets. The parity doublets of C_{60} simply provide a convenient tool for us to penetrate the maze of strong interactions.

This research was supported in part by the U.S. Department of Energy.

References

[1] H. W. Kroto *et al.*, Nature **318**, 162 (1985); W. Krätschmer *et al.*, Nature **347**, 354 (1990); H. Ajie *et al.*, J.Phys.Chem. **94**, 8630 (1990).

[2] A.F. Hebard *et al.*, Nature **350**, 600 (1991).

[3] M.J. Rosseinsky *et al.*, Phys.Rev.Lett. **66**, 2830 (1991).

[4] K. Holczer *et al.*, Science **252**, 1154 (1991).

[5] K. Holczer *et al.*, Phys.Rev.Lett. **67**, 271 (1991).

[6] G. Sparn *et al.*, Science **252**, 1829 (1991); R.M. Fleming *et al.*, Nature **352**, 787 (1991); O. Zhou *et al.*, Science **255**, 833 (1992).

[7] P.-M. Allemand *et al.*, Science **253**, 301 (1991).

[8] S. Chakravarty, M.P. Gelfand and S. Kivelson, Science **254**, 970 (1991), and UCLA preprint "Electronic Correlation Effects and Superconductivity in Doped Fullerenes".

[9] M. Hamermesh, *Group Theory and Its Application to Physical Problems* (Addison-Wesley Publishing Co., 1962).

[10] F. Albert Cotton, *Chemical Applications of Group Theory*, 2nd edition (Wiley-Interscience, 1963).

[11] R. Friedberg, T. D. Lee and H. C. Ren, Phys.Rev. **B46**, 14 150 (1992).

The Power of Duality –
Exact Results in 4D SUSY Field Theory

N. Seiberg

Department of Physics and Astronomy
Rutgers University
Piscataway, NJ 08855-0849, USA

and

Institute for Advanced Study
Princeton, NJ 08540, USA

Recently the vacuum structure of a large class of four dimensional (supersymmetric) quantum field theories was determined exactly. These theories exhibit a wide range of interesting new physical phenomena. One of the main new insights is the role of "electric-magnetic duality." In its simplest form it describes the long distance behavior of some strongly coupled, and hence complicated, "electric theories" in terms of weakly coupled "magnetic theories." This understanding sheds new light on confinement and the Higgs mechanism and uncovers new phases of four dimensional gauge theories. We review these developments and speculate on the outlook.

To appear in the Proc. of PASCOS 95 and in the Proc. of the Oskar Klein Lectures.

1. Introduction

Exact solutions play a crucial role in physics. It is often the case that a simple model exhibits the same phenomena which are also present in more complicated examples. The exact solution of the simple model then teaches us about more generic situations. For example, the exact solutions of the harmonic oscillator and of the hydrogen atom demonstrated many of the crucial aspects of quantum mechanics. Similarly, the Ising model was a useful laboratory in the study of statistical mechanics and quantum field theory. Many other exactly solvable two dimensional field theories like the Schwinger model and others have led to the understanding of many mechanisms in quantum field theory, which are also present in four dimensions.

The main point of this talk is to show how four dimensional supersymmetric quantum field theories can play a similar role as laboratories and testing grounds for ideas in more generic quantum field theories. This follows from the fact that these theories are more tractable than ordinary, non-supersymmetric theories, and many of their observables can be computed exactly. Nevertheless, it turns out that these theories exhibit explicit examples of various phenomena in quantum field theory. Some of them had been suggested before without an explicit realization and others are completely new. (For a brief review summarizing the understanding as of a year ago, see [1].)

Before continuing we would like to mention that supersymmetric four dimensional field theories also have two other applications:

1. Many physicists expect supersymmetry to be present in Nature, the reason being that it appears to be the leading candidate for solving the gauge hierarchy problem. If this is indeed the case, it is likely to be discovered experimentally in the next round of accelerators. Independent of the hierarchy problem, supersymmetry plays an important role in string theory as an interesting extension of our ideas about space and time. If Nature is indeed supersymmetric, understanding the dynamics of supersymmetric theories will have direct experimental applications. In particular, we will have to understand how supersymmetry is broken, i.e. why Nature is not exactly supersymmetric.
2. Witten discovered an interesting relation between supersymmetric field theories and four dimensional topology [2]. Using this relation, the exact solutions lead to a simplification of certain topological field theories and with that to advances in topology [3].

Even though these applications are important, here we will focus on the dynamical issues and will not discuss them.

The main insight that the study of these theories has taught us so far is the role of electric-magnetic duality in strongly coupled non-Abelian gauge theories[1]. Therefore, we will start our discussion in the next section by reviewing the duality in Abelian gauge theories, i.e. in electrodynamics. In section 3 we will present supersymmetric field theories and will outline how they are solved. This discussion will be rather heuristic. In section 4 we will summarize the results and will present the duality in non-Abelian theories. Finally in section 5 we will present our conclusions and a speculative outlook.

2. Duality in electrodynamics

2.1. The Coulomb phase

The simplest phase of electrodynamics is the Coulomb phase. It is characterized by massless photons which mediate a long range $\frac{1}{R}$ potential between external sources. In the absence of sources the relevant equations are Maxwell's equations in the vacuum

$$
\begin{aligned}
&\nabla \cdot E = 0\\
&\nabla \times B - \frac{\partial}{\partial t} E = 0\\
&\nabla \cdot B = 0\\
&\nabla \times E + \frac{\partial}{\partial t} B = 0
\end{aligned}
$$

(we have set the speed of light $c = 1$). Clearly, they are invariant under the duality transformation

$$
\begin{aligned}
E &\longrightarrow B\\
B &\longrightarrow -E
\end{aligned}
$$

which exchanges electric and magnetic fields.

If charged particles are added to the equations, the duality symmetry will be preserved only if both electric charges and magnetic monopoles are present. However, in Nature we

---

[1] The role of electric-magnetic duality in four dimensional quantum field theory was first suggested by Montonen and Olive [4]. Then, it became clear that the simplest version of their proposal is true only in $N = 4$ supersymmetric field theories [5] and in certain $N = 2$ supersymmetric theories [6]. Here we will discuss the extension of these ideas to $N = 1$ theories [7].

see electric charges but no magnetic monopole has been observed yet. This fact ruins the duality symmetry. We usually also ruin the symmetry by solving two of the equations by introducing the vector potential. Then, these equations are referred to as the Bianchi identities while the other two equations are the equations of motion. Had there also been magnetic monopoles, this would have been impossible.

Dirac was the first to study the possible existence of magnetic monopoles in the quantum theory. He derived the famous Dirac quantization condition which relates the electric charge e and the magnetic charge g:

$$eg = 2\pi$$

(we have set Planck's constant $\hbar = 1$). This relation has many important consequences. One of them is that since duality exchanges electric and magnetic fields, it also exchanges

$$e \quad \longleftrightarrow \quad g.$$

Since the product $eg = 2\pi$ is fixed, it relates weak coupling ($e \ll 1$) to strong coupling ($g \gg 1$). Therefore, even if we could perform such duality transformations in quantum electrodynamics, they would not be useful. Electrodynamics is weakly coupled because e is small. Expressing it in terms of "magnetic variables" will make it strongly coupled and therefore complicated.

However, one might hope that in other theories like QCD, such a duality transformation exists. If it does, it will map the underlying "electric" degrees of freedom of QCD, which are strongly coupled, to weakly coupled "magnetic" degrees of freedom. We would then have a weakly coupled, and therefore easily understandable, effective description of QCD.

2.2. The Higgs phase

When charged matter particles are present, electrodynamics can be in another phase – the superconducting or the Higgs phase. It is characterized by the condensation of a charged field ϕ

$$\langle\phi\rangle \neq 0.$$

This condensation creates a gap in the spectrum by making the photon massive. This phenomenon was first described in the context of superconductivity, where ϕ is the Cooper pair. It has since appeared in different systems including the weak interactions of particle physics where ϕ is the Higgs field.

The condensation of ϕ makes electric currents superconducting. Its effect on magnetic fields is known as the Meissner effect. Magnetic fields cannot penetrate the superconductor except in thin flux tubes. Therefore, when two magnetic monopoles (e.g. the ends of a long magnet) are inserted in a superconductor, the flux lines are not spread. Instead, a thin flux tube is formed between them. The energy stored in the flux tube is linear in its length and therefore the potential between two external magnetic monopoles is linear (as opposed to the $\frac{1}{R}$ potential outside the superconductor). Such a linear potential is known as a *confining* potential.

Mandelstam and 'tHooft considered the dual of this phenomenon: if instead of electric charges, magnetic monopoles condense, then magnetic currents are superconducting while electric charges are confined. Therefore, confinement is the dual of the Higgs mechanism. They suggested that confinement in QCD can be understood in a similar way by the condensation of color magnetic monopoles.

To summarize, we see that duality exchanges weak coupling and strong coupling. Therefore, it exchanges a description of the theory with small quantum fluctuations with a description with large quantum fluctuations. Similarly, it exchanges the weakly coupled phenomenon of the Higgs mechanism with the strong coupling phenomenon of confinement.

Such a transformation between variables which fluctuate rapidly and variables which are almost fixed is similar to a Fourier transform. When a coordinate is localized, its conjugate momentum fluctuates rapidly and vice versa. A duality transformation is like a Fourier transform between electric and magnetic variables.

It should be stressed, however, that except in simple cases (like electrodynamics without charges and some examples in two dimensional field theory) an explicit duality transformation is not known. It is not even known whether such a transformation exists at all. As we will show below, at least in supersymmetric theories such a transformation does exist (even though we do not have an explicit description of it).

3. The Dynamics of Supersymmetric Field Theories

In supersymmetric theories the elementary particles are in representations of supersymmetry. Every gauge boson, a gluon, is accompanied by a fermion, a gluino, and every matter fermion, a quark, is accompanied by a scalar, a squark. The theory is specified by a choice of a gauge group, which determines the coupling between the gluons, and a matter representation, which determines the coupling between the quarks and the gluons. For example, in QCD the gauge group is $SU(N_c)$ (N_c is the number of colors) and the matter representation is $N_f \times (\mathbf{N_c} + \overline{\mathbf{N}}_\mathbf{c})$ (N_f is the number of flavors).

An important object in these theories is the superpotential, $W(q)$. It is a holomorphic (independent of $\bar{q}$, the complex conjugate of q) gauge invariant function of the squarks, q. It determines many of the coupling constants and interactions in the theory including the Yukawa couplings of the quarks and the squarks and the scalar potential, $V(q, \bar{q})$. For example, a mass term for the quarks appears as a quadratic term in the superpotential.

The analysis of the classical theory starts by studying the minima of the scalar potential $V(q, \bar{q})$. It is often the case that the potential has many different degenerate minima. Then, the classical theory has many inequivalent ground states. Such a degeneracy between states which are not related by a symmetry is known as an "accidental degeneracy." Typically in field theory such an accidental degeneracy is lifted by quantum corrections. However, in supersymmetric theories, the degeneracy often persists in the quantum theory. Therefore, the quantum theory has a space of inequivalent vacua which can be labeled by the expectation values of the squarks $\langle q \rangle$.

This situation of many different ground states in the quantum theory is reminiscent of the situation when a symmetry is spontaneously broken. However, it should be stressed, that unlike that case, where the different ground states are related by a symmetry, here they are inequivalent. Physical observables vary from one vacuum to the other – they are functions of $\langle q \rangle$.

Our problem is to solve these theories not only as a function of all the parameters (like quark masses) but, for every value of the parameters, also as a function of the ground state $\langle q \rangle$.

As in Landau-Ginzburg theory, the best way to organize the information is in a low energy effective Lagrangian, L_{eff}. It includes only the low lying modes and describes their interactions. This effective theory is also specified by a gauge group and a matter representation. Since the theory is supersymmetric[2], the interactions of the light particles are characterized by a superpotential, W_{eff}.

The key observation is that this effective superpotential can often be determined exactly by imposing the following constraints [8]:

[2] We limit ourselves to theories where supersymmetry is not spontaneously broken.

1. In various limits of the parameter space or the space of ground states the theory is weakly coupled. In these limits $W_{\rm eff}$ can be determined approximately by weak coupling techniques (examples are the instanton calculations of [9-11]).
2. $W_{\rm eff}$ must respect all the symmetries in the problem. It is important to use also the constraints following from symmetries which are explicitly broken by the various coupling constants. Such symmetries lead to selection rules.
3. $W_{\rm eff}$ is holomorphic in the light fields and the parameters. This constraint is the crucial one which makes supersymmetric theories different from non-supersymmetric ones. The use of holomorphy in [8] generalized previous related ideas [12-16].

The superpotential determined in this way teaches us about the light particles and their interactions. This information determines the phase structure of the theory and the mechanisms for phase transitions.

By repeating this process for different theories, i.e. different gauge groups and matter representations, a rich spectrum of phenomena has been found. In the next section we will describe some of them.

4. Results – duality in non-Abelian theories

4.1. $SU(N_c)$ *with* $N_f \times (\mathbf{N_c} + \overline{\mathbf{N}}_\mathbf{c})$

Here the physical phenomena depend crucially on the number of massless quarks [9,17,15,18,7]:

I. $N_f \geq 3N_c$

In this range the theory is not asymptotically free. This means that, because of screening, the coupling constant becomes smaller at large distances. Therefore, the spectrum of the theory at large distance can be read off from the Lagrangian – it consists of the elementary quarks and gluons. The long distance behavior of the potential between external electric sources is of the form

$$V \sim \frac{1}{R \log R}.$$

The logarithm in the denominator shows that the interactions between the particles are weaker than in Coulomb theory. Therefore, we refer to this phase of the theory as a free electric phase.

We should add here that, strictly speaking, such a theory is not well defined as an interacting quantum field theory. However, it can be a consistent description of the low energy limit of another theory.

II. $\frac{3}{2}N_c < N_f < 3N_c$

In this range the theory is asymptotically free. This means that at short distance the coupling constant is small and it becomes larger at longer distances. However, in this regime [18,7] rather than growing to infinity, it reaches a finite value – a fixed point of the renormalization group.

Therefore, this is a non-trivial four dimensional conformal field theory. The elementary quarks and gluons are not confined there but appear as interacting massless particles. The potential between external electric sources behaves as

$$V \sim \frac{1}{R}$$

and therefore we refer to this phase of the theory as the non-Abelian Coulomb phase.

It turns out that there is an equivalent, "magnetic," description of the physics at this point [7]. It is based on the gauge group $SU(N_f - N_c)$, with N_f flavors of quarks and some gauge invariant fields. We will refer to this gauge group as the magnetic gauge group and to its quarks as magnetic quarks. This theory is also in a non-Abelian Coulomb phase because $\frac{3}{2}(N_f - N_c) < N_f < 3(N_f - N_c)$. The surprising fact is that its large distance behavior is identical to the large distance behavior of the original, "electric," $SU(N_c)$ theory. Note that the two theories have different gauge groups and different numbers of interacting particles. Nevertheless, they describe the same fixed point. In other words, there is no experimental way to determine whether the $\frac{1}{R}$ potential between external sources is mediated by the interacting electric or the interacting magnetic variables. Such a phenomenon of two different Lagrangians describing the same long distance physics is common in two dimensions and is known there as quantum equivalence. These four dimensional examples generalize the duality [4] in finite $N = 4$ supersymmetric theories [5] and in finite $N = 2$ theories [6] to asymptotically free $N = 1$ theories.

As N_f is reduced (e.g. by giving masses to some quarks and decoupling them) the electric theory becomes stronger – the fixed point of the renormalization group occurs at larger values of the coupling. Correspondingly, the magnetic theory becomes weaker. This can be understood by noting that as N_f is reduced, the magnetic gauge group becomes smaller. This happens by the Higgs mechanism in the magnetic theory [7].

III. $N_c + 2 \leq N_f \leq \frac{3}{2}N_c$

In this range the electric theory is very strongly coupled. However, since $3(N_f - N_c) \leq N_f$, the equivalent magnetic description based on the gauge group $SU(N_f - N_c)$ is not asymptotically free and it is weakly coupled at large distances. Therefore, the low energy spectrum of the theory consists of the particles in the dual magnetic Lagrangian [7]. These magnetic massless states are composites of the elementary electric degrees of freedom. The massless composite gauge bosons exhibit gauge invariance which is not visible in the underlying electric description. The theory generates new gauge invariance! We will return to this phenomenon in the conclusions.

This understanding allows us to determine the long distance behavior of the potential between magnetic sources:

$$V \sim \frac{1}{R \log R}.$$

Since the magnetic variables are free at long distance, we refer to this phase as a free magnetic phase.

IV. $N_f = N_c + 1, N_c$

As we continue to decouple quarks by giving them masses and thus reducing N_f, the magnetic gauge group is Higgsed more and more. Eventually it is completely broken and there are no massless gauge bosons. This complete Higgsing of the magnetic theory can be interpreted as complete confinement of the electric variables. We thus see an explicit realization of the ideas of Mandelstam and 'tHooft about confinement.

For $N_f = N_c + 1$ this confinement is not accompanied by chiral symmetry breaking, while for $N_f = N_c$ chiral symmetry is also broken [18]. In the electric language we describe the spectrum in terms of gauge invariant fields. These include massless mesons and baryons. In the magnetic language these are elementary fields. The idea that some of the composites in QCD, in particular the baryons, can be thought of as solitons was suggested in [19]. Here we see an explicit realization of a related idea – the baryons are magnetic monopoles composed of the elementary quarks and gluons.

V. $N_f < N_c$

In this range the theory of massless quarks has no ground state [9,15].

4.2. $SO(N_c)$ with $N_f \times \mathbf{N_c}$

In the $SU(N_c)$ theories there is no invariant distinction between Higgs and confinement [20]. This is not the case in theories based on $SO(N_c)$ with $N_f \times \mathbf{N_c}$ and therefore, they lead to a clearer picture of the dynamics. In particular, here the transition from the Higgs phase to the Confining phase occurs with a well defined phase transition.

Many of the results in these $SO(N_c)$ theories [9,21-23,7,24] are similar to the results in $SU(N_c)$ showing that these phenomena are generic. Here the duality map is

$$SO(N_c) \text{ with } N_f \times \mathbf{N_c} \longleftrightarrow SO(N_f - N_c + 4) \text{ with } N_f \times (\mathbf{N_f} - \mathbf{N_c} + \mathbf{4})$$

Let us consider three special cases of the duality:

1. For $N_c = 2$, $N_f = 0$ the map is

$$SO(2) \cong U(1) \longleftrightarrow SO(2) \cong U(1)$$

which is the ordinary duality of electrodynamics. Therefore, our duality is compatible with and generalizes this duality.

2. For $N_c = 3$, $N_f = 1$ the map is:

$$SO(3) \text{ with } \mathbf{3} \longleftrightarrow SO(2) \cong U(1) \text{ with } \mathbf{2}.$$

This theory was first analyzed in [22]. Here it is possible to understand the duality in more detail than in the more general case and in particular to identify the matter fields in the magnetic theory as magnetic monopoles. This theory has $N = 2$ supersymmetry but this did not play a role in our discussion. However, using the extra supersymmetry one can derive more exact results about the massive spectrum of the theory [22].

3. For $N_f = N_c - 2$ the map is

$$SO(N_c) \text{ with } (N_c - 2) \times \mathbf{N_c} \longleftrightarrow SO(2) \cong U(1) \text{ with } (N_c - 2) \times \mathbf{2}$$

which generalizes the previous example to situations without $N = 2$ supersymmetry. These three examples are simple because the magnetic theory is Abelian. However, in general the magnetic $SO(N_f - N_c + 4)$ gauge group is non-Abelian.

These $SO(N_c)$ theories also exhibit many new phenomena, which are not present in the $SU(N_c)$ examples. The most dramatic of them is oblique confinement [25,26], driven by the condensation of dyons (particles with both electric and magnetic charges). This phenomenon is best described by another equivalent theory – a dyonic theory. Therefore, these theories exhibit electric-magnetic-dyonic triality [24].

5. Conclusions

To conclude, supersymmetric field theories are tractable and many of their observables can be computed exactly[3]. The main dynamical lesson we learn is the role of electric-magnetic duality in non-Abelian gauge theories in four dimensions. This duality generalizes the duality in Maxwell's theory and in $N = 4$ [4] and certain $N = 2$ supersymmetric theories [6].

The magnetic degrees of freedom are related to the underlying electric degrees of freedom in a complicated (non-local) way. They are the effective degrees of freedom useful for describing the long distance behavior of the theory. These variables give a weak coupling description of strong coupling phenomena such as confinement.

Our analysis led us to find new phases of non-Abelian gauge theories, like the non-Abelian Coulomb phase with its quantum equivalence and the free magnetic phase with its massless composite gauge bosons.

Outlook

We would like to end by suggesting some future directions for research and speculate about them.

1. The exploration of supersymmetric models is far from complete. There are many models whose dynamics are not yet understood. It is likely that there are new phenomena in quantum field theory which can be uncovered here.
2. Of particular importance are chiral theories whose matter content is not in a real representation of the gauge group. Only a few of these have been analyzed [9,15,28]. These theories are also interesting as they can lead to dynamical supersymmetry breaking. Finding a nice model of dynamical supersymmetry breaking which is phenomenologically acceptable is an important challenge. For some recent work in this direction see [47].
3. An important question is to what extent these results are specific to supersymmetric theories. It would be very interesting to extend at least some of these ideas to non-supersymmetric theories and to find the various phases and the mechanisms for the phase transitions without supersymmetry. One way such a study can proceed is by perturbing a supersymmetric theory whose solution is known by soft breaking terms.

[3] Although we did not discuss them here, we would like to point out that many other examples were studied [9,15,27,21,6,23,7,28-46] exhibiting many new interesting phenomena.

When these terms are small they do not affect the dynamics significantly and the non-supersymmetric perturbed theory is qualitatively similar to the unperturbed theory (for a recent attempt in this direction see [48]). This makes it clear that the phenomena we found can be present in non-supersymmetric theories as well. Alternatively, one can start with ordinary non-supersymmetric QCD with a large enough number of flavors N_f, where a non-trivial fixed point of the renormalization group exists [49] and the theory is in a non-Abelian Coulomb phase. It is not yet known for which values of N_f this phase exists. It is possible that there is a dual magnetic description of this phase and perhaps even a non-Abelian free magnetic phase exists for some values of N_f. One might be tempted to speculate that perhaps the confinement of ordinary QCD can be described as a Higgs phenomenon in these variables. Then, perhaps some of the massive particles in the spectrum of QCD (like the ρ meson or the A_1) can be identified as "Higgsed magnetic gluons."

4. The duality points at a big gap in our current understanding of gauge theories. We do not have an explicit transformation relating the underlying electric degrees of freedom to their magnetic counterparts. Finding such an explicit transformation will be extremely interesting. If such a transformation does not exist within the standard framework of local quantum field theory, perhaps a reformulation of quantum field theory will be needed.
5. The duality reflects new gauge invariance, which is not obvious in the fundamental description of the theory. In hindsight this is not surprising because gauge symmetry is not a symmetry. It is merely a redundancy in the description. Perhaps we should conclude that gauge symmetries might not be fundamental! They might only appear as long distance artifacts of our description of the theory. If so, perhaps some of the gauge symmetries of the standard model or even general relativity are similarly long distance artifacts. Then, the corresponding gauge particles are the "magnetic" degrees of freedom of more elementary "electric" variables. In order not to violate the theorem of [50], the underlying theory cannot have these symmetries as global symmetries. In particular, this can be the case for gravity only if the underlying theory is topological.
6. Recently, there has been enormous interest in dualities and non-perturbative effects in string theory [51]. Some of the phenomena found in string theory are similar to and generalize those found in field theory. It would be nice to use some of the techniques which turned out to be useful in field theory in string theory as well.

Finally we should note that there are obvious relations between these directions. For example, progress in understanding the origin of the duality can help the study of supersymmetric chiral theories and lead to phenomenological theories of supersymmetry breaking.

Acknowledgments

We would like to thank T. Banks, D. Kutasov, R. Leigh, M.R. Plesser, P. Pouliot, S. Shenker, M. Strassler and especially K. Intriligator and E. Witten for many helpful discussions. This work was supported in part by DOE grant #DE-FG05-90ER40559.

References

[1] N. Seiberg, The Power of Holomorphy – Exact Results in 4D SUSY Field Theories. To appear in the Proc. of PASCOS 94. hep-th/9408013, RU-94-64, IASSNS-HEP-94/57

[2] E. Witten, Comm. Math. Phys. 117 (1988) 353.

[3] E. Witten, hep-th/9411102, Math. Res. Lett. 1 (1994) 769.

[4] C. Montonen and D. Olive, Phys. Lett. 72B (1977) 117; P. Goddard, J. Nuyts and D. Olive, Nucl. Phys. B125 (1977) 1.

[5] H. Osborn, Phys. Lett. 83B (1979) 321; A. Sen, hep-th/9402032, Phys. Lett. 329B (1994) 217; C. Vafa and E. Witten, hep-th/9408074, Nucl. Phys. B432 (1994) 3.

[6] N. Seiberg and E. Witten, hep-th/9408099, Nucl. Phys. B431 (1994) 484.

[7] N. Seiberg, hep-th/9411149, Nucl. Phys. B435 (1995) 129.

[8] N. Seiberg, hep-ph/9309335, Phys. Lett. 318B (1993) 469.

[9] I. Affleck, M. Dine and N. Seiberg, Nucl. Phys. B241 (1984) 493; Nucl. Phys. B256 (1985) 557.

[10] N. Seiberg, Phys. Lett. 206B (1988) 75.

[11] D. Finnell and P. Pouliot, RU-95-14, SLAC-PUB-95-6768, hep-th/9503115.

[12] E. Witten, Nucl. Phys. B268 (1986) 79.

[13] M. Dine and N. Seiberg, Phys. Rev. Lett. 57 (1986) 2625.

[14] M.A. Shifman and A. I. Vainshtein, Nucl. Phys. B277 (1986) 456; Nucl. Phys. B359 (1991) 571.

[15] D. Amati, K. Konishi, Y. Meurice, G.C. Rossi and G. Veneziano, Phys. Rep. 162 (1988) 169 and references therein.

[16] J. Polchinski and N. Seiberg, (1988) unpublished.

[17] V.A. Novikov, M.A. Shifman, A. I. Vainshtein and V. I. Zakharov, Nucl. Phys. B223 (1983) 445; Nucl. Phys. B229 (1983) 381; Nucl. Phys. B260 (1985) 157.

[18] N. Seiberg, hep-th/9402044, Phys. Rev. D49 (1994) 6857.

[19] T.H.R. Skyrme, Proc.Roy.Soc. A260 (1961) 127; E. Witten, Nucl. Phys. B160 (1979) 57; Nucl. Phys. B223 (1983) 422; Nucl. Phys. B223 (1983) 433.

[20] T. Banks, E. Rabinovici, Nucl. Phys. B160 (1979) 349; E. Fradkin and S. Shenker, Phys. Rev. D19 (1979) 3682.

[21] K. Intriligator, R.G. Leigh and N. Seiberg, hep-th/9403198, Phys. Rev. D50 (1994) 1092; K. Intriligator, hep-th/9407106, Phys. Lett. 336B (1994) 409.

[22] N. Seiberg and E. Witten, hep-th/9407087, Nucl. Phys. B426 (1994) 19.

[23] K. Intriligator and N. Seiberg, hep-th/9408155, Nucl. Phys. B431 (1994) 551.

[24] K. Intriligator and N. Seiberg, RU–95–3, hep-th/9503179; RU-95-40, IASSNS-HEP-95/48, to appear in the Proc. of Strings 95.

[25] G. 'tHooft, Nucl. Phys. B190 (1981) 455.

[26] J. Cardy and E. Rabinovici, Nucl. Phys. B205 (1982) 1; J. Cardy, Nucl. Phys. B205 (1982) 17.

[27] A.Yu. Morozov, M.A. Olshansetsky and M.A. Shifman, Nucl. Phys. B304 (1988) 291.

[28] K. Intriligator, N. Seiberg and S. Shenker, hep-ph/9410203, Phys. Lett. 342B (1995) 152.

[29] A. Klemm, W. Lerche, S. Theisen and S. Yankielowicz, hep-th/9411048, Phys. Lett. 344B (1995) 169; hep-th/9412158.

[30] P. Argyres and A. Faraggi, hep-th/9411057.

[31] O. Aharony, hep-th/9502013, TAUP-2232-95.

[32] D. Kutasov, hep-th/9503086, EFI-95-11.

[33] R. Leigh and M. Strassler, RU-95-2, hep-th/9503121.

[34] U. Danielsson and B. Sundborg, USITP-95-06, UUITP-4/95, hep-th/9504102.

[35] M.R. Douglas and S.H. Shenker, RU-95-12, RU-95-12, hep-th/9503163.

[36] S. Elitzur, A Forge, A. Giveon and E. Rabinovici, RI-4-95 hep-th/9504080.

[37] O. Aharony, J. Sonnenschein and S. Yankielowicz, TAUP–2246–95, CERN-TH/95–91, hep-th/9504113.

[38] D. Kutasov and A. Schwimmer, EFI–95–20, WIS/4/95, hep-th/9505004.

[39] K. Intriligator and P. Pouliot, RU-95-23, hep-th/9505006.

[40] K. Intriligator, RU–95–27, hep-th/9505051.

[41] P.C. Argyres and M.R. Douglas, RU-95-31, hep-th/9505062.

[42] M. Berkooz, RU-95-29, hep-th/9505067.

[43] R. Leigh and M. Strassler, hep-th/9505088, RU-95-30.

[44] A. Hanany and Y. Oz, TAUP-2248-95, WIS-95/19, hep-th/9505075.

[45] P.C. Argyres, M.R. Plesser and A. Shapere, IASSNS-HEP-95/32, UK-HEP/95-06, hep-th/9505100.

[46] K. Intriligator, R. Leigh and M. Strassler, RU-95-38, to appear.

[47] M. Dine and A.E. Nelson and Y. Shirman, hep-ph/9408384, Phys. Rev. D51 (1995) 1362.

[48] N. Evans, S.D.H. Hsu, M. Schwetz, hep-th/9503186, YCTP-P8-95.

[49] T. Banks and A. Zaks, Nucl. Phys. B196 (1982) 189.

[50] S. Weinberg and E. Witten, Phys. Lett. 96B (1980) 59.

[51] At the current rate of progress any list of references will be enormous and will immediately become outdated.

String theory as a universal language

A.M. Polyakov

Joseph Henry Laboratories
Princeton University
Princeton, New Jersey 08544

Abstract

This article, based on the Klein lecture, contains some new results and new speculations on various topics. They include discussion of open strings in the AdS space, unusual features of D-branes, conformal gauge theories in higher dimensions. We also comment on the infrared screening of the cosmological constant and on the "brane worlds"

1 Introduction

String theory is a beautiful and dangerous subject. On one hand it is a top achievement of theoretical physics exploiting the most advanced and daring methods. On the other — without a guidance from the experiment it can easily degenerate into a collection of baroque curiosities, some kind of modern alchemy looking for philosopher's stone.

This danger can be somewhat reduced if we try to study string theory in connection with some concrete physical problem and then extrapolate the gained experience to the Planck domain unreachable by experiments. This is a well established strategy in theoretical physics. For example one can learn about the Cherenkov radiation while studying supersonic aerodynamics. And usually there is the "back-reaction": the technical progress at the frontier turns out to be helpful in solving the old problems. Thus, it is conceivable that string theory will provide us with the language for the future theoretical physics.

In this lecture I will examine a number of problems in which the language of string theory is appropriate and effective. We begin with the problem of quark confinement. The task here is to find the string description of the color-electric flux lines emerging in QCD. Recently there has been a considerable progress in this field. Various aspects of it have been reviewed in [1, 2]. I shall not review again these developments and instead concentrate on new results. After that we will discuss some general features of D-branes, conformal gauge theories in higher dimensions, and speculations concerning the cosmological constant.

2 The image of gluons and zigzag symmetry

As was explained in the above references, a string theory, needed to describe gauge fields in 4 dimensions, must be formulated in the 5d space with the metric

$$ds^2 = d\varphi^2 + a^2(\varphi)d\vec{x}^2 \tag{1}$$

This curved 5d space is a natural habitat for the color-electric flux lines. If apart from the pure gauge fields there are some matter fields in the theory, they must correspond to extra degrees of freedom on the world sheet. In some cases these extra degrees of freedom can be balanced so that the field theory β−function is equal to zero. These cases are the easiest ones, since the conformal symmetry of the field theory requires conformal symmetry of the string background and determines it completely:

$$a(\varphi) \sim e^{\alpha\varphi},$$

where α is some constant. After an obvious change of variables the metric takes the form

$$ds^2 = \sqrt{\lambda} y^{-2}(dy^2 + d\vec{x}^2), \tag{2}$$

where λ is related to the coupling constant of field theory. When $\lambda \gg 1$ the curvature of this 5d space is small and the 2d sigma model describing the string in the above background is weakly coupled. This greatly simplifies the analyses and we will concentrate on this case. Our aim in this section will be to demonstrate that open strings in this background have some very unusual properties, allowing to identify them with gluons.

Before starting let us recall that at present we have two possible approaches to the question of field-string correspondence. None of them is fully justified but both have certain heuristic power. In the first approach one begins with a stack of D-branes describing a gauge theory and then replace the stack by its gravitational background. In the second approach one doesn't introduce D-branes and starts directly with the sigma model action, adjusting the background so that the boundary states of this string describe the gauge theory. The key principle here is the zigzag symmetry. This is a requirement that these boundary states consist of vector gluons (and matter fields, if present) and nothing else.

The situation is very unusual. Normally we have an infinite tower of states in both open and closed string sectors. Here we need a string theory in which the closed string sector contains an infinite number of states, while the open sector has a finite number of the field-theoretic states. Our first task will be to explore how this is possible.

To set the stage, let us remember how open strings are treated in the standard case [3]. One begins with the action

$$S = \frac{1}{2}\int_D (\partial x)^2 + i\int_{\partial D} A_\mu dx_\mu, \tag{3}$$

where D is a unit disc, ∂D — its boundary, and A_μ is a vector condensate of the open string states. The possible fields A_μ are determined from the condition that the functional integral

$$Z[A] = \int Dx\, e^{-S} \tag{4}$$

is conformally invariant. The explicit form of this condition is derived by splitting

$$x = c + z, \tag{5}$$

where c is a slow variable, while z is fast, and integrating out z. Conformal invariance requires vanishing of the divergent counterterms and that restricts the background fields. It is convenient to integrate first the fields inside the disk with the fields at the boundary being fixed. That gives the standard boundary action

$$S_B = \frac{1}{2}\int \frac{dudv}{(u-v)^2}(x(u)-x(v))^2 + i\int A_\mu dx_\mu. \tag{6}$$

We see that x_μ are the Gaussian fields with the correlation function (in the momentum space)

$$< x_\mu(p)x_\nu(-p) >\sim \delta_{\mu\nu} \mid p \mid^{-1}. \tag{7}$$

Expanding the second term in z we obtain

$$S_B \approx \frac{1}{2}\sum \mid p \mid z(p)z(-p) + \int \nabla_\lambda F_{\sigma\mu}(c)\frac{dc_\mu}{ds} z_\lambda(s) z_\sigma(s) ds. \tag{8}$$

Using the fact that

$$< z_\lambda(s) z_\sigma(s) >\sim \delta_{\lambda\sigma}\int^\Lambda \frac{dp}{\mid p \mid}, \tag{9}$$

where Λ is an ultraviolet cut-off, we obtain as condition that the divergence cancel (in this approximation)

$$\nabla_\lambda F_{\lambda\mu} = 0. \tag{10}$$

This is the on-shell condition for the massless string mode. Qualitatively the same treatment is applicable to the massive states as well. In this case one perturbs S_B with the operator

$$\Delta S_B = \int ds \Psi(x(s))(\dot{x}^2(s)h^{-2}(s))^n h(s), \tag{11}$$

where Ψ is the scalar massive mode at the level $2n$ and $h(s)$ is the boundary metric on the world sheet needed for the general covariance of this expression. Conformal invariance of this perturbation means that the $h(s)$ dependence must cancel. The cancellation occurs between the explicit h -dependence in the above formula and the factors coming from the quantum fluctuations of $x(s)$. These factors appear because in the covariant theory the cut-off is always accompanied by the boundary metric

$$h(s)(\Delta s)^2_{\min} = a^2, \tag{12}$$

$$\Lambda^2 = \frac{1}{(\Delta s)^2_{\min}} = \frac{1}{a^2}h(s), \tag{13}$$

where a is an invariant cut-off.

In the one loop approximation (which is not, strictly speaking, applicable here, but gives a correct qualitative picture) we have

$$\partial^2\Psi - M_n^2\Psi = 0, \tag{14}$$

$$M_n^2 \sim n. \tag{15}$$

This is the on-shell condition for the massive string mode and it was obtained, let us stress it again, from the cancellation between the classical and quantum $h(s)$ dependence

Now we are ready to attack the AdS case. Let us consider the string action in this background

$$S = \sqrt{\lambda}\int_D \frac{(\partial x_\mu)^2 + (\partial y)^2}{y^2} + ..., \tag{16}$$

where we dropped all fermionic and RR terms. This is legitimate in the WKB limit $\lambda \gg 1$ which we will study in this section. To find the counterterms we must once again calculate the boundary action and $< z_\lambda(s) z_\mu(s) >$. It is not as easy as in the previous case, but this well-defined mathematical problem was solved in [4, 5]. The answer has the following form

$$S_{cl} = \sqrt{\lambda} \int ds_1 ds_2 \kappa_{\mu\nu}(s_1, s_2) z_\mu(s_1) z_\nu(s_2). \tag{17}$$

After introducing variables $s = \frac{s_1+s_2}{2}$ and $\sigma = s_1 - s_2$ and taking the Fourier transform with respect to σ we obtain the following asymptotic for the kernel in the mixed representation

$$\kappa_{\mu\nu}(p,s) \underset{p\to\infty}{\approx} \frac{|p|^3}{(c'(s))^2}[3\frac{c'_\mu c'_\nu}{(c')^2} - \delta_{\mu\nu}]. \tag{18}$$

¿From this it follows that

$$< z_\mu(s) z_\nu(s) > \propto \int^\infty \frac{dp}{|p|^3} < \infty. \tag{19}$$

The remarkable feature of this answer is that it implies that *there is no quantum ultraviolet divergences on the world sheet.* Hence, if we add to the action the background fields

$$\Delta S \sim \int A_\mu dx_\mu + \int \Psi(x(s))(x'(s))^2 (h(s))^{-1} ds + \ldots \tag{20}$$

and treat it in the one loop approximation, we come to the following conclusions. First of all, as far as the A-term is concerned, it is finite for any $A_\mu(x)$ and thus describes the off-shell gluons. This situation is in the sharp contrast with the standard case in which conformal invariance implied the on-shell condition.

Now let us examine the massive mode (11). The only quantum dependence on the cut-off comes from

$$< z'_\mu(s) z'_\nu(s) > \sim \int \frac{dp}{|p|} \tag{21}$$

As a result we obtain the following counterterm

$$\Delta S \sim \int \Psi(x(s))(x'(s))^2 (\frac{\log h(s)}{h(s)}) ds \tag{22}$$

We see that the only way to keep the theory conformally invariant in this approximation is to set $\Psi = 0$. There is also a possibility that at some fixed value of λ the $h(s)$ dependence will go away. However it is impossible to cancel it by a suitable on-shell condition. All this happens because, due to (19), the kinetic energy for the Ψ-term is not generated.

There is one more massless mode in the AdS string which requires a special treatment. Let us examine

$$\Delta S = \int \Phi(x(s))\partial_\perp y(s)\, ds, \tag{23}$$

where $\partial_\perp$ is a normal derivative at the boundary of the world sheet (which lies at infinity of the AdS space). In the more general case of $\mathrm{AdS_p \times S_q}$ we have also a perturbation

$$\Delta S = \int \Phi^i(x) n^i(s) \partial_\perp y ds. \tag{24}$$

Here we must remember that the string action is finite only if [6, 5]

$$(\partial_\perp y)^2 = (x')^2 . \tag{25}$$

It is easy to see that when we substitute the decomposition (5) into this formula we get the logarithmic divergence (21) once again. We come to the conclusion that the above perturbation is not conformal and must not be present at the boundary. However in the case of (24) there is also a logarithmic term, coming from the fluctuations of $n^i(s)$. In the presence of space-time supersymmetry these two divergences must cancel since the masslessness of the scalar fields is protected by the SUSY. Otherwise, keeping the scalar fields massless requires a special fine tuning of the background. It would be interesting to clarify the corresponding mechanism.

Another interesting problem for the future is the fate of the open string tachyon in the AdS space. So far we assumed that it is excluded by the GSO projection. But in the purely bosonic string it may lead to some interesting effects via Sen mechanism [7].

We come to the following conclusion concerning the spectrum of the boundary states in the AdS-like background. It consists of a few modes which would have been massless in the flat case. The infinite tower of the massive states can not reach the boundary. The above finite set of states must be associated with the fields of the field theory under consideration.

The full justification of this assertion requires the analyses of the Schwinger-Dyson equations of the Yang-Mills theory. It is still absent, and we give some heuristic arguments instead. The loop equation, expressing the Schwinger-Dyson equations in terms of loops has the form

$$\widehat{L}(s)W(C) = W * W \tag{26}$$

where $W(C)$ is the Wilson loop and $\widehat{L}$ is the loop laplacian and the right hand side comes from the self-intersecting contours. In recent papers [4, 5] we analyzed the action of the loop Laplacian in the AdS space. It was shown that at least in the WKB approximation and in the four -dimensional space-time we have a highly non-trivial relation

$$\widehat{L}(s) \sim T_{\perp||}(s) \tag{27}$$

where $T_{\perp||}(s)$ is a component of the world sheet energy-momentum tensor at the boundary. When substituted inside the string functional integral the energy momentum tensor receives contributions from the degenerate metrics only. These metrics describe a pinched disk, that is two discs joined at a point. The corresponding amplitude is saturated by the allowed boundary operators inserted at this point. That gives the equation (26) provided that the boundary operators of the string are the same as the fields of the field theory. Much work is still needed to make this argument completely precise.

3 The D-brane picture

An alternative way to understand gauge fields-strings duality is based on the D-brane approach. It is less general than the sigma model approach described above, but in the supersymmetric cases it provides us with an attractive visual picture. The logic of this method is based on the fundamental conjecture that D-branes can be described as some particular solitons in the closed string sector. One of the strongest arguments in favor of this conjecture is that both D-branes and solitons have the same symmetries and are sources of the same RR fields [8]. The gauge fields -strings dualities then follows from the D-branes - solitons duality in the limit $\alpha' \to 0$. On the D-brane side only the massless gauge field modes of the open string survive in this limit. On the string theory side we have a near horizon limit [1] of the soliton metric [9] , given by (1) . These two theories must be equivalent, if the basic D-brane conjecture is correct.

The connection with the sigma model approach of the previous section follows from the following argument. First of all the closed string background is the same in both cases. As for the open strings, we placed their ends at the boundary of the AdS space where $a^2(\varphi) \to \infty$. That means that the effective slope of this strings behaves as

$$\alpha'_{open} \sim a^{-2}(\varphi) \to 0 \tag{28}$$

and thus only the massless modes are present. We said that the D-brane approach is less general, because in the non-supersymmetric cases there could exist solitons with the required boundary behavior, which are not describable by any combination of D-branes in the flat space.

This fact is related to another often overlooked subtlety. The 3-brane soliton has a metric [9]

$$ds^2 = H^{-\frac{1}{2}}(r)(dx)^2 + H^{\frac{1}{2}}(r)(dy)^2 \tag{29}$$

$$r^2 = y^2; \quad H(r) = 1 + \frac{L^4}{r^4} \tag{30}$$

It is often assumed that this metric is an extremum of the action

$$S = S_{bulk} + S_{BI} \tag{31}$$

where the first term contains the modes of the closed string, while the second is the Born -Infeld action localized on the brane. In the equations of motion the second term will give a delta function of the transverse coordinates.

Would it be the case, where the 3-brane is located? From (30) it is clear that the singularity of the metric is located in the complex domain $r^4 = -L^4$. Let us try to understand the significance of this fact from the string-theoretic point of view. Consider a string diagram describing the D-brane world volume in the arbitrary order in $\lambda = g_s N$. It is represented by a disc with an arbitrary number of holes. At each boundary one imposes the Dirichlet conditions for the transverse coordinates. An important feature of this diagram is that it is finite. This follows from the fact that the only source of divergences in string theory are tadpoles and for 3-branes their contribution is proportional to the integral $\int \frac{d^6k}{k^2}$ where k is the transverse momentum. This expression is infrared finite (which is of course very well known). Thus D-branes in the flat space are described by the well defined string amplitudes. But that contradicts the common wisdom, that one must determine the background from the action (31) , because the *flat space is not a solution,* once the Born-Infeld term is added. More over if we try to deform the flat space, the above disc with holes will loose its conformal invariance.

Let us analyze this apparent paradox. It is related to the fact on a sphere conformal invariance is equivalent to the absence of tadpoles since for any (1,1) vertex operator we have $< V >_{sphere} = 0$. However on a disc this is not true, the conformal symmetry doesn't forbid the expectation values of vertex operators. On a disc conformal symmetry and the absence of tadpoles are two different conditions. Which one should we use?

If we denote the bulk couplings by λ and the boundary couplings by μ we can construct three different objects, the bulk central charge $c(\lambda)$, the "boundary entropy" [10] $b(\lambda, \mu)$ and the effective action generating the S-matrix, $S(\lambda, \mu) = c(\lambda) + b(\lambda, \mu)$. To ensure conformal invariance we must have

$$\frac{\partial c}{\partial \lambda} = 0 \tag{32}$$

$$\frac{\partial b}{\partial \mu} = 0 \tag{33}$$

This does not coincide in general with the "no tadpole condition"

$$\frac{\partial S}{\partial \lambda} = \frac{\partial S}{\partial \mu} = 0 \tag{34}$$

In the case of 3-branes the paradox is resolved in an interesting way. The metric (30) has a horizon at $r = 0$. When we go to the Euclidean signature the horizon, as usual, shrinks to a non-singular point. As a result we have a metric which solves the equation (32) and has no trace of the D-brane singularity in it! The paradox is pushed under the horizon.

The conclusion of this discussion is as following. We have two dual and different descriptions of the D-brane amplitudes. In the first description we calculate the amplitudes of a disc with holes *in the flat space.* In this description it is simply inconsistent to introduce the background fields generated by D-branes.

In the second description we forget about the D-branes and study a non-singular closed string soliton. The D-brane conjecture implies that we must get the same answers in these two cases. The situation is analogous to the one we have in the sine-gordon theory, which admits two dual descriptions, either in terms of solitons or in terms of elementary fermions, but not both.

Let us touch briefly another consequence of these considerations. When minimizing the action (31) one can find a solution which is singular on the 3-brane and is AdS-space outside of it [11]. These solutions are known to "localize" gravitons on the brane and are the basis of the popular "brane-world" scenarios. It is clear that for the string-theoretic branes this is not a physical solution because the world volume does not contain gravity (being described by the open strings). As we argued above, there must be a horizon, not a singularity. Technically this happens because in string theory the Born-Infeld action is corrected with the Einstein term $\sim \int R\sqrt{g}d^{p+1}x$, coming from the finite thickness of the brane. It can be shown that the coefficient in front of this term (which is fully determined by string theory) is tuned so that the localization is destroyed. There are no worlds on D-branes. Of course if one compactifies the ambient transverse space, the 4d graviton reappears by the Kaluza-Klein mechanism.

4 Conformal gauge theories in higher dimensions

Although our main goal is to find a string theoretic description of the asymptotically free theories, conformal cases are not without interest. They are easier and can be used as a testing ground for the new methods. In this section we briefly discuss conformal bosonic gauge theories in various dimensions [2]. The background in these cases is just the AdS space. We have to perform the non-chiral GSO projection in order to eliminate the boundary tachyon (it would add an instability to the field theory under consideration; we do not consider here an interesting possibility that this instability resolves in some new phase).

The GSO projection in the non-critical string is slightly unusual. Let us consider first d=5 (corresponding to the d=4 gauge theory). In this case we have 4 standard NSR fermions ψ_μ on the world sheet and also a partner of the Liouville field ψ_5. For the former we can use the standard spin fields defined by the OPE

$$\psi_\mu \times \Sigma_A \sim (\gamma_\mu)_{AB}\, \Sigma_B \tag{35}$$

where we use the usual 4×4 Dirac matrices. The spinor Σ_A can be split into spinors $\Sigma_A^{\pm}$ with positive and negative chiralities. It is easy to check that the OPE for them have the structure

$$\Sigma^{\pm} \times \Sigma^{\pm} \quad \sim \quad (\psi)^{[even]} \tag{36}$$

$$\Sigma^{\pm} \times \Sigma^{\mp} \quad \sim \quad (\psi)^{[odd]} \tag{37}$$

The symbols on the RHS mean the products of even/odd number of the NSR fermions. Notice that this structure is the opposite to the one in 10 dimension. In order to obtain the spin operators in 5d we have to introduce the Ising order and disorder operators , σ and μ , related to ψ_5. These operators are non-holomorphic and correspond to RR-states. Their OPE have the structure

$$\sigma \times \sigma \quad \sim \quad (\psi_5)^{[even]} \tag{38}$$

$$\sigma \times \mu \quad \sim \quad (\psi_5)^{[odd]} \tag{39}$$

Using these relations we obtain the following GSO projected RR spin operator

$$\Sigma = \begin{pmatrix} \sigma\Sigma^+\overline{\Sigma^+} & \mu\Sigma^+\overline{\Sigma^-} \\ \mu\Sigma^-\overline{\Sigma^+} & \sigma\Sigma^-\overline{\Sigma^-} \end{pmatrix} \tag{40}$$

It has the property

$$\Sigma \times \Sigma \sim (\psi)^{[even]} \tag{41}$$

needed for the non-chiral GSO projection, which consists of dropping all operators with the odd number of fermions. Notice also that the non-chiral picture changing operator has even number of fermions. The RR matrix Σ has 16 elements. We can now write down the full string action in the AdS_5 space. It has the form

$$S = S_B + S_F + S_{RR} + S_{ghost} \tag{42}$$

where S_B is given by (16), S_F by

$$S_F = \int d^2\xi[\overline{\psi}_M \nabla \psi_M + \frac{1}{\sqrt{\lambda}}(\overline{\psi_M}\gamma_\mu \psi_N)^2] \tag{43}$$

where $M = 1, ...5$,and one have to use the standard spin-connection projected from AdS on the world sheet in the Dirac operator. So far we are describing the usual action of the sigma model with N=1 supersymmetry on the world sheet. The unusual part is the RR term given by

$$S_{RR} = f \int d^2\xi Tr(\gamma_5 \Sigma) e^{-\frac{\phi}{2}} \tag{44}$$

Here $e^{-\frac{\phi}{2}}$ is a spin operator for the bosonic ghost [12] and f is a coupling constant (which in one loop approximation is equal to 1).

I believe that this model can be exactly solved, although it has not been done yet. A promising approach to this solution may be based on the non-abelian bosonization [13] in which the fermions ψ_M are replaced by the orthogonal matrix Ω_{MN} with the WZNW Lagrangian. In this case the RR term is simply a trace of this matrix in the spinor representation. This formalism lies in the middle between the NSR and the Green-Schwartz approaches and hopefully will be useful. Meanwhile we will have to be content with the one loop estimates which are justified in some special cases listed below and help to get a qualitative picture in general. In this approach one begins with the effective action

$$S = \int d^d x \sqrt{G} e^{\Phi} \left[\frac{d-10}{2} - R - (\nabla\Phi)^2 \right] + \int d^d x F_d^2 \sqrt{G} \tag{45}$$

Here F_d is the RR d-form and $F_d^2 = G^{A_1 A_1'} ... G^{A_d A_d'} F_{A_1 ... A_d} F_{A_1' ... A_d'}$; the form F_d will be assumed to be proportional to the volume form . The dilaton field Φ is normalized so that $e^{\Phi} = g_s^{-2}$, where g_s is the string coupling constant. Conformal cases involve either constant curvature solutions of the equations of motion or the products of the manifolds with constant curvatures. The dilaton in these cases is also a constant. Such an ansatz is very easy to analyze. Let us begin with the single AdS$_d$ space. Consider the variation of the metric which preserves the constancy of the curvature

$$\delta G_{AB} = \varepsilon G_{AB} \tag{46}$$

$$\delta R = \delta(G^{AB} R_{AB}) = -\varepsilon R \tag{47}$$

$$\delta\Phi = const \tag{48}$$

That immediately gives the relations

$$\frac{d-10}{2} - R = 0 \tag{49}$$

$$e^{\Phi}(1-\frac{d}{2})R - \frac{d}{2}F_d^2 = 0 \tag{50}$$

If the assume that the flux of the RR field is equal to N, we get $F_d^2 = N^{2\cdot}$. If we introduce the coupling constant $\lambda = g_s N = g_{YM}^2 N$, we obtain the background AdS solution [2] with

$$\mid R \mid \sim \lambda^2 \sim 10 - d \tag{51}$$

We can trust this solution if $d = 10 - \epsilon$, in which case the curvatures and the RR fields are small and the above one loop approximation is justified. According to the discussion in the preceding sections, this solution must describe the Yang-Mills theory, perhaps with one adjoint scalar, in the space with dimension $9-\epsilon$. We conclude that this bosonic higher dimensional gauge theory has a conformally invariant fixed point! It may be worth mentioning that this is not atypical for non-renormalizable theories to have such fixed points. For example a nonlinear sigma model in dimension higher than two, where it is non-renormalizable, does have a conformal critical point at which the phase transition to ferromagnetic phase takes place. However it is hard to say up to what values of ϵ we can extrapolate this result.

These considerations allow for several generalizations. First of all we can consider products of spaces with constant curvatures by the same method. Take for example the space $\mathrm{AdS}_p \times S_q$ with curvatures $R_1$and R_2 and $d = p+q$. To get the equations of motion in this case it is sufficient to consider the variations

$$\delta G_{ab} = \varepsilon_1 G_{ab} \tag{52}$$

$$\delta G_{ij} = \varepsilon_2 G_{ij} \tag{53}$$

where the first part refers to AdS and the second to the sphere. A simple calculation gives the equations

$$\frac{d-10}{2} - R_1 - R_2 = 0 \tag{54}$$

$$(1-\frac{2}{p})R_1 + \frac{10-d}{2} = -e^{-\Phi}F_d^2 \tag{55}$$

$$(1-\frac{2}{q})R_2 + \frac{10-d}{2} = e^{-\Phi}F_d^2 \tag{56}$$

$$F_d^2 = \lambda^2 R_2^q \tag{57}$$

Here we assume that the RR flux is permeating the AdS component of space only, being given by the volume form. The last equation follows from the normalization condition of this flux and the extra factor proportional to the volume of S_q in the action. Solving this equations we get

$$R_1 = -(\frac{10-p-q}{2})\frac{p(q+2)}{p-q} \tag{58}$$

$$R_2 = (\frac{10-p-q}{2})\frac{q(p+2)}{p-q} \tag{59}$$

This solution describes bosonic gauge theories with $q+1$ adjoint bosons; again it can be trusted if the curvatures are small. Another generalization is related to the fact that strictly speaking we must include the closed string tachyon in our considerations. It was shown in [14] that there exists an interesting mechanism for the tachyon condensation, following from its couplings to the RR fields. It is easy to include the constant tachyon field in our action and to show that it doesn't change our results in the small curvature limit. According to [14], in the critical case $d = 10$ the tachyon leads to the running coupling constants. In the non-critical case there is also a conformal option, described above, in which the tachyon condenses to a constant value.

Finally let us describe the reasons to believe that the conformal solutions can be extrapolated to non-small curvatures and thus the sigma model (42) has a conformal fixed point. The first two terms in (45) are the expansion of the sigma model central charge. When the couplings are not small, we have to replace

$$\frac{d-10}{2} - R \Rightarrow \frac{c(R)-10}{2} \tag{60}$$

where $c(R)$ is the the central charge of the 2d sigma model with the target space having a curvature R. It decreases , according to Zamolodchikov, along the renormalization group trajectory. We can call this the second law of the renormalization group. Let us conjecture that there is also a *third law*

$$c(R) \rightarrow 0 \quad (R \rightarrow +\infty) \tag{61}$$

$$c(R) \rightarrow \infty \quad (R \rightarrow -\infty) \tag{62}$$

The first property follows from the fact that usually the sigma models with positive curvature develop a mass gap and thus there are no degrees of freedom contributing to the central charge. The second equation is harder to justify; we know only that $c(R)$ is increasing in the direction of negative curvature.

With this properties and with some general form of the RR terms it is possible to see that the conformal solution to the equations of motion, obtained by the variations (46), continue to exist when ϵ is not small. Of course this is not a good way to explore these solutions. Instead one must construct the conformal algebra for the sigma model action(42). This has not been done yet.

5 Infrared screening of the cosmological constant and other speculations

In this section we will discuss some speculative approaches to the problems of vacuum energy and space-time singularities. I shall try to revive some old ideas [15], adding some additional thoughts. The motivation to do that comes from the remarkable recent observational findings indicating that the cosmological constant is non- zero, and its scale is defined by the size of the universe (meaning the Hubble constant). These results seem very natural from the point of view advocated in [15], according to which there is an almost complete screening of the cosmological constant due to the infrared fluctuations of the gravitational field. This phenomenon is analogous to the complete screening of electric charge in quantum electrodynamics found by Landau, Abrikosov and Khalatnikov and nicknamed "Moscow zero". Here we will try to argue in favor for another "zero" of this kind -that of the cosmological constant.

Let us consider at first the Einstein action

$$S = -\int (R - 2\Lambda_0)\sqrt{g} d^4x \tag{63}$$

Here Λ_0 is a bare cosmological constant which is assumed to be defined by the Planck scale. It is clear from the form of the action that if we consider the infrared fluctuations of the metric (with the wave length much larger then the Planck scale) their interaction will be dominated by the second term in this formula since it doesn't contain derivatives. To get some qualitative understanding of the phenomenon, let us consider conformally flat fluctuations of the metric

$$g_{\mu\nu} = \varphi^2 \delta_{\mu\nu} \tag{64}$$

The action takes the form

$$S = -\int [\frac{1}{2}(\partial\varphi)^2 - \Lambda_0\varphi^4] d^4x \tag{65}$$

It has the well known feature of non-positivity. The way to treat it was suggested in [16] and we will accept it, although it doesn't have good physical justification. To use S in the functional integral we will simply analytically continue $\varphi \Rightarrow i\varphi$; after that the action takes the form

$$S = \int [\frac{1}{2}(\partial\varphi)^2 + \Lambda_0\varphi^4] d^4x \tag{66}$$

The infrared fluctuations of φ are relevant and lead to the screening of Λ_0

$$\Lambda \sim \frac{1}{\log(M_{pl}L)} \tag{67}$$

where L is an infrared cut-off. More generally we could represent the metric in the form

$$g_{\mu\nu} = \varphi^2 h_{\mu\nu};\quad \det(h_{\mu\nu}) = 1 \tag{68}$$

It is not known how to treat the unimodular part of the metric. We can only hope that it will not undo the infrared screening although can change it. Also the screening (67) with $L \sim H^{-1}$ (where H is the Hubble constant) is not strong enough to explain the fact that $\Lambda \sim H^2$. It is not impossible that the two problems cure each other. When we have several relevant degrees of freedom, the renormalization group equations governing the L dependence of various coupling constants including Λ may have a powerlike asymptotic (in contrast with (67)). Such examples exist, starting from the cases with two independent coupling constants. Thus the infrared limit of the Einstein action (perhaps with the dilaton added) may be described by a conformal field theory, giving the cosmological constant defined by the Hubble scale. The renormalization group should take us from Planck to Hubble.

Even if this fantasy is realized, we have to resolve another puzzle. It is certainly unacceptable to have a large cosmological constant in the early Universe, since it will damage the theory of nucleosynthesis. At the first glance it seems to create a serious problem for the screening theory,because when the universe is relatively small, the screening is small also and the cosmological constant is large. The way out of this problem is to conjecture that in the radiation dominated universe the infrared cut-off is provided by the curvature of space-time, while in the matter-dominated era it begins to depend on other quantities characterizing the size of the universe. If this is the case, in the early universe we get the screening law $\Lambda \sim R$ instead of $\Lambda \sim H^2$. Substituting it in the Einstein action we find that at this stage the infrared mechanism simply renormalizes the Newton constant and thus is unobservable. In the matter dominated era the effective Λ begin to depend on other things (like the Friedman warp factor a) and that can easily give the observed acceleration of the universe. Thus the change in the infrared screening must be related to a trace of the energy momentum tensor. We can say that in the correct theory the cosmological constant vanishes without the trace. To be more precise, it is disguised as a Newton constant, until the trace of the energy-momentum tensor reveals its true identity. The above picture have some remote resemblance to the scenario suggested recently in [17].

In spite of the obvious gaps in these arguments, they give a very natural way of relating the cosmological constant to the size of the universe and thus are worth developing. Perhaps the AdS/CFT correspondence will be of some use for this purpose. The main technical problem in testing these ideas is the unusual $\varphi-$ dependent kinetic energy of the h-field.

We can also notice that the above scenario can explain the dimensionality of space-time. Indeed, if this dimensionality is larger then four, the infrared effects are small, the cosmological constant- large, and we end up in the universe of the Planck size, which is not much fun.

This work was supported in part by the NSF grant PHY9802484.

REFERENCES

[1] O. Aharony, S.Gubser, J. Maldacena, H. Ooguri, Phys. Rep. 323 (2000) 183

[2] A. Polyakov J. Mod. Phys. A14 (1999) 645, hep-th/9809057

[3] C. Callan et al. Nucl. Phys. B308 (1988) 221

[4] A. Polyakov, V. Rychkov hep-th/0002106

[5] A. Polyakov, V. Rychkov hep-th/0005173

[6] N. Drukker, D. Gross, H. Ooguri Phys. Rev.D60 (199) 125, hep-th/9904191

[7] A. Sen JHEP 9912027, hep-th/9911116

[8] J. Polchinski Phys. Rev. Lett. 75 (1995) 4724, hep-th/9510017

[9] G. Horowitz, A. Strominger Nucl. Phys. B360 (1991) 197

[10] I. Affleck, A. Ludwig Phys.Rev.Lett. 67 (1991) 161

[11] L. Randall, R. Sundrum Phys. Rev. Lett. 83 (1999) 4690

[12] D. Friedan, E. Martinec, S. Shenker Nucl. Phys.B271 (1986) 93

[13] E. Witten Comm. Math. Phys.92 (1984) 455

[14] Ig. Klebanov A. Tseytlin Nucl. Phys.B547 (1999) hep-th/9812089

[15] A. Polyakov Sov. Phys. Uspekhi 25 (1982) 187

[16] G. Gibbons, S. Hawking Phys.Rev.D15 (1977) 2725

[17] C. Armendariz-Picon, V. Mukhanov, P. Steinhardt, astro-ph/0004139

The Cosmological Tests

P. J. E. Peebles

Joseph Henry Laboratories, Princeton University, Princeton, NJ 08544

Abstract. Recent observational advances have considerably improved the cosmological tests, adding to the lines of evidence, and showing that some issues under discussion just a few years ago may now be considered resolved or irrelevant. Other issues remain, however, and await resolution before the great program of testing the relativistic Friedmann-Lemaître model, that commenced in the 1930s, may at last be considered complete.

1. Introduction

The search for a well-founded physical cosmology is ambitious, to say the least, and the heavy reliance on philosophy not encouraging. When Einstein (1917) adopted the assumption that the universe is close to homogeneous and isotropic, it was quite contrary to the available astronomical evidence; Klein (1966) was right to seek alternatives. But the observations now strongly support Einstein's homogeneity. Here is a case where philosophy led us to an aspect of physical reality, even though we don't know how to interpret the philosophy. Other aesthetic considerations have been less successful. The steady-state cosmology and the Einstein-de Sitter relativistic cosmology are elegant, but inconsistent with the observational evidence we have now.

The program of empirical tests of cosmology has been a productive research activity for seven decades. The results have greatly narrowed the options for a viable cosmology, and show that the relativistic Friedmann-Lemaître model passes impressively demanding checks. My survey of the state of the tests four years ago, in a Klein lecture, was organized around Table 1. I discuss here the provenance of this table, and how it would have to be revised to fit the present situation.

The table is crowded, in part because I tried to refer to the main open issues; there were lots of them. Some are resolved, but an updated table would be even more crowded, to reflect the considerable advances in developing new lines of evidence. This greatly enlarges the checks for consistency that are the key to establishing any element of physical science. Abundant checks are particularly important here, because astronomical evidence is limited, by what Nature chooses to show us and by our natural optimism in interpreting it.

Figure 1. THE COSMOLOGICAL TESTS—A SCORECARD

Test	Einstein-de Sitter	$\Omega = 0.2 \pm 0.15$ $R^{-2} = 0$	$\Omega = 0.2 \pm 0.15$ $\Lambda = 0$	Issues for Physics	Issues for Astronomy & Astrophysics
1a. Dynamics on scales $\lesssim 10$ Mpc	X	√	√	The inverse square law; nature of the	What's in the voids? Where and when
1b. Dynamics on larger scales	X?	√	√	dark mass & its effective Jeans length	did galaxies form? What is $\Omega_{\rm eff}$ *vs.* scale?
2a. World time $t(z)$ & ages of stars & elements	X??	√	√	Radial displacement-redshift relation	It's the distance scale, stupid
2b. Redshift-angular size relation				$dl/dz = c\,dt/dz$; angular size	Evolution: galaxy merging & accretion;
2c. Redshift-magnitude relation	√	?	√	distance $r(z)$; Liouville relation	star populations & SNe; ISM & dust;
2d. Counts: dN/dm, dN/dz, $dN = f(m,z)dm\,dz$	X??	?	√	$i \propto (1+z)^{-4}$; gravitational	AGN central engines; ionizing background
2e. Lensing by galaxies & clusters of galaxies	√	?	√	deflection of light	Mass structures, dust in lensing objects
3a. Baryon mass fractions in clusters of galaxies	X	√	√	BBNS; nature of the primeval departure	The structures of clusters of galaxies;
3b. Cluster structures $\rho(r)$ & number density at $z \ll 1$				from homogeneity: near scale-invariant	evolution of cluster luminosity, baryon &
3c. Cluster evolution $N(\sigma, L, M_{\rm B}, z)$	X	√?	√	adiabatic Gaussian fluctuations? strings?	mass functions; angular & redshift surveys;
3d. $\xi_{gg}(r,z)$, ξ_{cc}, ξ_{cg}, $\xi_{g\rho}$, $\xi_{\rho\rho}$, $\Gamma = \Omega h$, superclusters, ...	X??	√	√	non-Gaussian &/or isocurvature?; models	large-scale peculiar motions & lensing
3e. Redshifts of assembly of Lα clouds, galaxies, QSOs, ...				for nonlinear & non-gravitational processes	How did galaxies form in the first generations?
4. Baryon density $\Omega_{\rm B}$ & the origin of light elements				N_ν, m_ν, μ_ν, $s/n_{\rm B}$; B-fields; new physics	What is the present baryon density $\Omega_{\rm B}$?
5. 3 K CBR spectrum, anisotropy, polarization				All of the above	Effects of reionization & interactions at low z

P. J. E. Peebles, June 1997 √ = conditional pass; **X** = conditional fail

2. The Classical Tests

2.1. The Physics of the Friedmann-Lemaître Model

The second to the last column in the table summarizes elements of the physics of the relativistic Friedmann-Lemaître cosmological model we want to test.

The starting assumption follows Einstein in taking the observable universe to be close to homogeneous and isotropic. This agrees with the isotropy of the radiation backgrounds and of counts of sources observed at wavelengths ranging from radio to gamma rays. Isotropy allows a universe that is inhomogeneous but spherically symmetric about a point very close to us. I think it is not overly optimistic to consider this picture unlikely, but in any case it is subject to the cosmological tests, through its effect on the redshift-magnitude relation, for example.

It will be recalled that, if a homogeneous spacetime is described by a single metric tensor, the line element can be written in the Robertson-Walker form,

$$ds^2 = dt^2 - a(t)^2 \left(\frac{dr^2}{1 \pm r^2/R^2} + r^2(d\theta^2 + \sin^2\theta d\phi^2) \right). \tag{1}$$

This general expression contains one constant, that measures the curvature of sections of constant world time t, and one function of t, the expansion parameter $a(t)$. Under conventional local physics the de Broglie wavelength of a freely moving particle varies as $\lambda \propto a(t)$. In effect, the expansion of the universe stretches the wavelength. The stretching of the wavelengths of observed freely propagating electromagnetic radiation is measured by the redshift, z, in terms of the ratio of the observed wavelength of a spectral feature to the wavelength measured at rest at the source,

$$1 + z = \frac{\lambda_{\text{observed}}}{\lambda_{\text{emitted}}} = \frac{a(t_{\text{observed}})}{a(t_{\text{emitted}})}. \tag{2}$$

At small redshift the difference δt of world times at emission and detection of the light from a galaxy is relatively small, the physical distance between emitter and observer is close to $r = c\delta t$, and the rate of increase of the proper distance is

$$v = cz = H_o r, \qquad H_o = \dot{a}(t_o)/a(t_o), \tag{3}$$

evaluated at the present epoch, t_o. This is Hubble's law for the general recession of the nebulae.

Under conventional local physics Liouville's theorem applies. It says an object at redshift z with radiation surface brightness i_e, as measured by an observer at rest at the object, has observed surface brightness (integrated over all wavelengths)

$$i_o = i_e(1 + z)^{-4}. \tag{4}$$

Two powers of the expansion factor can be ascribed to aberration, one to the effect of the redshift on the energy of each photon, and one to the effect of time dilation on the rate of reception of photons. In a static "tired light" cosmology one might expect only the decreasing photon energy, which would imply

$$i_o = i_e(1 + z)^{-1}. \tag{5}$$

Measurements of surface brightnesses of galaxies as functions of redshift thus can in principle distinguish these expanding and static models (Hubble & Tolman 1935).

The same surface brightness relations apply to the 3 K thermal background radiation (the CBR). Under equation (4) (and generalized to the surface brightness per frequency interval), a thermal blackbody spectrum remains thermal, the temperature varying as $T \propto a(t)^{-1}$, when the universe is optically thin. The universe now is optically thin at the Hubble length at CBR wavelengths. Thus we can imagine the CBR was thermalized at high redshift, when the universe was hot, dense, and optically thick. Since no one has seen how to account for the thermal CBR spectrum under equation (5), the CBR is strong evidence for the expansion of the universe. But since our imaginations are limited, the check by the application of the Hubble-Tolman test to galaxy surface brightnesses is well motivated (Sandage 1992).

The constant R^{-2} and function $a(t)$ in equation (1) are measurable in principle, by the redshift-dependence of counts of objects, their angular sizes, and their ages relative to the present. The metric theory on which these measurements are based is testable from consistency: there are more observable functions than theoretical ones.

The more practical goal of the cosmological tests is to over-constrain the parameters in the relativistic equation for $a(t)$,

$$\begin{aligned}\left(\frac{\dot{a}}{a}\right)^2 &= \frac{8}{3}\pi G\rho_t \pm \frac{1}{a^2R^2} \\ &= H_o^2[\Omega_m(1+z)^3 + \Omega_\Lambda + (1-\Omega_m-\Omega_\Lambda)(1+z)^2]. \qquad (6)\end{aligned}$$

The total energy density, ρ_t, is the time-time part of the stress-energy tensor. The second line assumes ρ_t is a sum of low pressure matter, with energy density that varies as $\rho_m \propto a(t)^{-3} \propto (1+z)^3$, and a nearly constant component that acts like Einstein's cosmological constant Λ. These terms, and the curvature term, are parametrized by their present contributions to the square of the expansion rate, where Hubble's constant is defined by equation (3).

2.2. Applications of the Tests

When I assembled Table 1 the magnificent program of application of the redshift-magnitude relation to type Ia supernovae was just getting underway, with the somewhat mixed preliminary results entered in line 2c (Perlmutter et al. 1997). Now the supernovae measurements clearly point to low Ω_m (Reiss 2000 and references therein), consistent with most other results entered in the table.

The density parameter Ω_m inferred from the application of equation (6), as in the redshift-magnitude relation, can be compared to what is derived from the dynamics of peculiar motions of gas and stars, and from the observed growth of mass concentrations with decreasing redshift. The latter is assumed to reflect the theoretical prediction that the expanding universe is gravitationally unstable. Lines 1a and 1b show estimates of the density parameter Ω_m based on dynamical interpretations of measurements of peculiar velocities (relative to the uniform expansion of Hubble's law) on relatively small and large scales. Some of the latter indicated $\Omega_m \sim 1$. A constraint from the evolution of clustering is entered in line 3c. The overall picture was pretty clear then, and now seems well established: within the Friedmann-Lemaître model the mass that clusters with the galaxies almost certainly is well below the Einstein-de Sitter case $\Omega_m = 1$. I discuss the issue of how much mass might be in the voids between the concentrations of observed galaxies and gas clouds in §4.1.

All these ideas were under discussion, in terms we could recognize, in the 1930s. The entry in line 2e is based on the prediction of gravitational lensing. That was well known in the 1930s, but the recognition that it provides a cosmological test is more recent. The entry refers to the multiple imaging of quasars by foreground galaxies (Fukugita & Turner 1991). The straightforward reading of the evidence from this strong lensing still favors small Λ, but with broad error bars, and it does not yet seriously constrain Ω_m (Helbig 2000). Weak lensing — the distortion of galaxy images by clustered foreground mass — has been detected; the inferred surface mass densities indicate low Ω_m (Mellier et al. 2001), again consistent with most of the other constraints.

The other considerations in lines 1 through 4 are tighter, but the situation is not greatly different from what is indicated in the table and reviewed in Lasenby, Jones & Wilkinson (2000).

3. The Paradigm Shift to Low Mass Density

Einstein and de Sitter (1932) argued that the case $\Omega_m = 1$, $\Omega_\Lambda = 0$, is a reasonable working model: the low pressure matter term in equation (6) is the only component one could be sure is present, and the Λ and space curvature terms are not logically required in a relativistic expanding universe.

Now everyone agrees that this Einstein-de Sitter model is the elegant case, because it has no characteristic time to compare to the epoch at which we have come on the scene. This, with the perception that the Einstein-de Sitter model offers the most natural fit to the inflation scenario for the very early universe, led to a near consensus in the early 1990s that our universe almost certainly is Einstein-de Sitter.

The arguments make sense, but not the conclusion. For the reasons discussed in §4.1, it seems to me exceedingly difficult to reconcile $\Omega_m = 1$ with the dynamical evidence.

The paradigm has shifted, to a low density cosmologically flat universe, with

$$\Omega_m = 0.25 \pm 0.10, \qquad \Omega_\Lambda = 1 - \Omega_m. \tag{7}$$

There still are useful cautionary discussions (Rowan-Robinson 2000), but the community generally has settled on these numbers. The main driver was not the dynamical evidence, but rather the observational fit to the adiabatic cold dark matter (CDM) model for structure formation (Ostriker & Steinhardt 1995).

I have mixed feelings about this. The low mass density model certainly makes sense from the point of view of dynamics (Bahcall et al. 2000; Peebles, Shaya & Tully 2000; and references therein). It is a beautiful fit to the new evidence from weak lensing and the SNeIa redshift-magnitude relation. But the paradigm shift was driven by a model for structure formation, and the set of assumptions in the model must be added to the list to be checked to complete the cosmological tests. I am driven by aspects of the CDM model that make me feel uneasy, as discussed next.

4. Issues of Structure Formation

A decade ago, at least five models for the origin of galaxies and their clustered spatial distribution were under active discussion (Peebles & Silk 1990). Five years later the community had settled on the adiabatic CDM model. That was in part because simple versions of the competing models were shown to fail, and in part because the CDM model was seen to be successful enough to be worthy of close analysis. But the universe is a complicated place: it would hardly be surprising to learn that several of the processes under discussion in 1990 prove to be significant dynamical actors, maybe along with things we haven't even thought of yet. That motivated my possibly overwrought lists of issues in lines 3a to 3e in the table.

I spent a lot of time devising alternatives to the CDM model. My feeling was that such a simple picture, that we hit on so early in the search for ideas on how structure formed, could easily fail, and it would be prudent to have backups. Each of my alternatives was ruled out by the inexorable advance of the measurements, mainly of the power spectrum of fluctuations in the temperature of the CBR. The details can be traced back through Hu & Peebles (2000). The experience makes me all the more deeply impressed by the dramatic success of the CDM model in relating observationally acceptable cosmological parameters to the measured temperature fluctuation spectrum (Hu et al 2000 and references therein).

This interpretation is not unique. McGaugh (2000) presents a useful though not yet completely developed alternative, that assumes there is no nonbaryonic dark matter — $\Omega_{\rm baryon} = \Omega_m \sim 0.04$ — and assumes a modification to the gravitational inverse square law on large scales — following Milgrom (1983) — drives structure formation. But the broad success of the CDM model makes a strong case that this is a good approximation to what happened as matter and radiation decoupled at redshift $z \sim 1000$. An update of Table 1 would considerably enlarge entry 5.

If the CDM model really is the right interpretation of the CBR anisotropy it leaves considerable room for adjustment of the details. I turn now to two issues buried in the table that might motivate a critical examination of details.

4.1. Voids

The issue of what is in the voids defined by the concentrations of observed galaxies and gas clouds is discussed at length in Peebles (2001); here is a summary of the main points.

The familiar textbook, optically selected, galaxies are strongly clustered, leaving large regions — voids — where the number density of galaxies is well below the cosmic mean. Galaxies with low gas content and little evidence of ongoing star formation prefer dense regions; gas-rich galaxies prefer the lower ambient density near the edges of voids. This is the morphology-density correlation.

The CDM model predicts that the morphology-density correlation extends to the voids, where the morphological mix swings to favor dark galaxies. But the observations require this swing to be close to discontinuous. A substantial astronomical literature documents the tendency of galaxies of all known types — dwarfs, irregulars, star forming, low surface brightness, and purely gaseous — to avoid the same void regions. There are some galaxies in voids, but they are not all that unusual, apart from the tendency for greater gas content.

The natural interpretation of these phenomena is that gravity has emptied the voids of most galaxies of all types, and with them drained away most of the low pressure mass. This is is not allowed in the Einstein-de Sitter model. If the mass corresponding to $\Omega_m = 1$ were clustered with the galaxies the gravitational accelerations would be expected to produce peculiar velocities well in excess of what is observed. That is, if $\Omega_m = 1$ most of the mass would have to be in the voids, and the morphology-density correlation would have to include the curious discontinuous swing to a mix dominated by dark galaxies in voids. That is why I put so much weight on the dynamical evidence for low Ω_m.

At $\Omega_m = 0.25 \pm 0.10$ (eq. [7]) the mass fraction in voids can be as small as the galaxy fraction. That would neatly remove the discontinuity. But gravity does not empty the voids in numerical simulations of the low density CDM model. In the simulations massive dark mass halos that seem to be suitable homes for ordinary optically selected galaxies form in concentrations. This is good. But spreading away from these concentrations are dark mass halos that are too small for ordinary galaxies, but seem to be capable of developing into dwarfs or irregulars. This is contrary to the observations.

The consensus in the theoretical community is that the predicted dark mass clumps in the voids need not be a problem, because we don't know how galaxies form, how to make the connection between dark mass halos in a simulation and galaxies in the real world. The point is valid, but we have some guidance, from what is observed. Here is an example.

The Local Group of galaxies contains two large spirals, our Milky Way and the somewhat more massive Andromeda Nebula. There many smaller galaxies, most tightly clustered around the two spirals. But some half dozen irregular galaxies, similar to the Magellanic clouds, are on the outskirts of the group. These irregulars have small velocities relative to the Local Group. Since they are not near either of the large galaxies they are not likely to have been spawned by tidal tails or other nonlinear process. Since they are at ambient densities close to the cosmic mean their first substantial star populations would have formed under conditions not very different from the voids at the same epoch. In short, these objects seem to prove by their existence that observable galaxies can form under conditions similar to the voids in CDM simulations. Why are such galaxies so rare in the voids?

4.2. The Epoch of Galaxy Formation

Numerical simulations of the CDM model indicate galaxies were assembled relatively recently, at redshift $z \sim 1$ (eg. Cen & Ostriker 2000). For definiteness in explaining what bothers me about this I adopt the density parameter in equation (7) and Hubble parameter $H_o = 70$ km s^{-1} Mpc^{-1}.

The mass in the central luminous parts of a spiral galaxy is dominated by stars. The outer parts are thought to be dominated by nonbaryonic dark matter. The circular velocity v_c of a particle gravitationally bound in a circular orbit in the galaxy varies only slowly with the radius of the orbit, and there is not a pronounced change in v_c at the transition between the luminous inner part and the dark outer part. The value of the mean mass density $\rho(< r)$ averaged within a sphere of radius r centered on the galaxy relative to the cosmic mean mass density, $\bar{\rho}$, is

$$\frac{\rho(< r)}{\bar{\rho}} = \frac{2}{\Omega_m}\left(\frac{v_c}{H_o r}\right)^2 \sim 3 \times 10^5, \quad \text{at} \quad r = 15 \text{ kpc}. \tag{8}$$

At this radius the mass of the typical spiral is thought to be dominated by nonbaryonic dark matter. Why is the dark mass density so large? Options are that

1. at formation the dark mass collapsed by a large factor,
2. massive halos formed by the merging of smaller dense clumps, that formed at high redshift, when the mean mass density was large, or
3. massive halos themselves were assembled at high redshift.

To avoid confusion let us pause to consider the distinction between interpretations of large density contrasts in the luminous baryonic central regions and in the dark halo of a galaxy. If the baryons and dark matter were well mixed at high redshift, the baryon-dominated central parts of the galaxies would have to have been the result of settling of the baryons relative to the dark matter. Gnedin, Norman, and Ostriker (2000) present a numerical simulation that demonstrates dissipative settling of the baryons to satisfactory stellar bulges. The result is attractive — and hardly surprising since gaseous baryons tend to dissipatively settle — but does not address the issue at hand: how did the dark matter halos that are thought to be made of dissipationless matter get to be so dense?

We have one guide from the great clusters of galaxies. The cluster mass is thought to be dominated by nonbaryonic matter. A typical line-of-sight velocity dispersion is $\sigma = 750$ km s^{-1}. The mean mass density averaged within the Abell radius, $r_{\rm A} = 2$ Mpc, relative to the cosmic mean, is

$$\frac{\rho(< r_A)}{\bar{\rho}} = \frac{4}{\Omega_m}\left(\frac{\sigma}{H_o r_{\rm A}}\right)^2 \sim 300. \qquad (9)$$

Clusters tend to be clumpy at the Abell radius, apparently still relaxing to statistical equilibrium after the last major mergers, but they are thought to be close to dynamic equilibrium, gravity balanced by streaming motions of the galaxies and mass. This argues against the first of the above ideas: here are dark matter concentrations that have relaxed to dynamical support at density contrast well below the dark halo of a galaxy (eq. [8]).

The second idea to consider is that a dense dark matter halo is assembled at low redshift by the merging of a collection of dense lower mass halos that formed earlier. This is what happens in numerical simulations of the CDM model. Sometimes cited as an example in Nature is the projected merging of the two Local Group spirals, the Milky Way and the Andromeda Nebula. They are 750 kpc apart, and moving together at 100 km s^{-1}. If they moved to a direct hit they would merge in another Hubble time. But that would require either wonderfully close to radial motion or dynamical drag sufficient to eliminate the relative orbital angular momentum. Orbit computations indicate masses that would be contained in $\rho \propto r^{-2}$ dark halos truncated at $r \sim 200$ kpc, which seems small for dissipation of the orbital angular momentum. The computations suggest the transverse relative velocity is comparable to the radial component (Peebles, Shaya & Tully 2000). That would say the next perigalacticon will be at a separation ~ 300 kpc, not favorable for merging. Thus I suspect the Local Group spirals will remain distinct elements of the galaxy clustering hierarchy well beyond one present Hubble time. If the Local Group is gravitationally bound and remains isolated the two spirals must eventually merge, but not on the time scale of the late galaxy formation picture.

My doubts are reenforced by the failure to observe precursors of galaxies. Galaxy spheroids — elliptical galaxies and the bulges in spirals that look like ellipticals — are dominated by old stars. Thus it is thought that if present-day spheroids were assembled at $z \sim 1$ it would have been by the merging of star clusters. These star clusters might have been observable at redshift $z > 1$, as a strongly clustered population, but they are not.

That leaves the third idea, early galaxy assembly. I am not aware of any conflict with what is observed at $z < 1$. The observations of what happened at higher redshift are rich, growing, and under debate.

5. Concluding Remarks

If the models for cosmology and structure formation on which Table 1 is based could be taken as given, the only uncertainties being the astronomy, the constraints on the cosmological parameters would be clear. The long list of evidence for $\Omega_m = 0.25 \pm 0.10$, in the table and the new results from the SNeIa redshift-magnitude relation and weak lensing, abundantly demonstrates that we live in a low density universe. The CBR demonstrates space sections are flat. Since Ω_m is small there has to be a term in the stress-energy tensor that acts like Einstein's cosmological constant. The SNeIa redshift-magnitude result favors a low density flat universe over low density with open geometry ($\Lambda = 0$), at about three standard deviations. That alone is not compelling, considering the hazards of astronomy, but it is an impressive check of what the CBR anisotropy says.

But we should remember that all this depends on models we are supposed to be testing. The dynamical estimates of Ω_m in lines 1a and 1b assume the inverse square law of gravity. That is appropriate, because it follows from the relativistic cosmology we are testing. We have a check on this aspect of the theory, from consistency with other observations whose theoretical interpretations depend on Ω_m in other ways. The CDM model fitted to observed large-scale structure requires a value of Ω_m that agrees with dynamics. This elegant result was an early driver for the adoption of the low density CDM model. But we can't use it as evidence for both the CDM model and the inverse square law; we must turn to other measures. We have two beautiful new results, from weak lensing and the redshift-magnitude relation, that agree with $\Omega_m \sim 0.25$. The latter does not exclude $\Omega_m = \Omega_{\mathrm{baryons}} \sim 0.04$; maybe MOND accounts for flat $v_c(r)$ but does not affect equation (6) (McGaugh 2000). And if we modified local Newtonian dynamics we might want to modify the physics of the gravitational deflection of light.

There are alternative fits to the CBR anisotropy, with new physics (McGaugh 2000), or conventional physics and an arguably desperate model for early structure formation (Peebles, Seager & Hu 2000). They certainly look a lot less elegant than conventional general relativity theory with the CDM model, but we've changed our ideas of elegance before.

In §4 I reviewed two issues in structure formation that I think challenge the CDM model. They may in fact only illustrate the difficulty of interpreting observations of complex systems. It's just possible that they will lead us to some radical adjustment of the models for structure formation and/or cosmology. I don't give much weight to this, because it would mean the model led us to the right Ω_m for the wrong reason. Relatively fine adjustments are easier to imagine, of course. With them we must be prepared for fine adjustments of the constraints on parameters such as Λ.

This is quite a tangled web. Progress in applying the many tests, including the mapping the CBR temperature and polarization, will be followed with close attention.

We have an impressive case for the Friedmann-Lemaître cosmology, from the successful fit to the CBR anisotropy and the consistency of the evidence for $\Omega_m \sim 0.25$ from a broad range of physics and astronomy. But the cosmological tests certainly are not complete and unambiguous, and since they depend on astronomy the program is not likely to be closed by one critical measurement. Instead, we should expect a continued heavy accumulation of evidence, whose weight will at last unambiguously compel acceptance. We are seeing the accumulation; we all look forward to the outcome.

Acknowledgments. This work was supported in part by the US National Science Foundation.

References

Bahcall, N. A., Cen, R., Davé, R., Ostriker, J. P. & Yu, Q. 2000, astro-ph/0002310

Cen, R. & Ostriker, J. P. 2000, ApJ, 538, 83

Einstein, A. 1917, S.-B. Preuss. Akad. Wiss. 142

Einstein, A. & de Sitter, W. 1932, Proc NAS, 18, 213

Fukugita, M. & Turner, E. L. 1991, MNRAS, 253, 99

Gneddin, N. Y., Norman, M. L. & Ostriker, J. P. 2000, astro-ph/9912563

Helbig, P. 2000, astro-ph/0011031

Hu, W., Fukugita, M., Zaldarriaga, M., & Tegmark, M. 2000, astro-ph/0006436

Hu, W. & Peebles, P. J. E. 2000, ApJ, 528, L61

Hubble, E. & Tolman, R. C. 1935, ApJ, 82. 302

Klein, O. 1966, Nature, 211, 1337

Lasenby, A., Jones, A. W. & Wilkinson, A. 2000, New Cosmological Data and the Values of the Cosmological Parameters, IAU Symposium 201

McGaugh, S. S. 2000, ApJ, 541, L33

Mellier et al. 2001, astro-ph/0101130

Milgrom, M. 1983, ApJ, 270, 371

Ostriker, J. P. & Steinhardt, P. J. 1995, Nature, 377, 600

Peebles, P. J. E. 2001, astro-ph/0101127

Peebles, P. J. E., Seager, S. & Hu, W. 2000, ApJ, 539, L1

Peebles, P. J. E., Shaya, E. J. & Tully, R. B. 2000, astro-ph/0010480

Peebles, P. J. E. & Silk, J. 1990, Nature, 346, 233

Perlmutter, S. et al. 1997, ApJ, 483, 565

Reiss, A. G. 2000, PASP, 112, 1248

Rowan-Robinson, M. 2000, astro-ph/0012026

Sandage, A. R. 1992, Physica Scripta, T43, 22

Anti-de Sitter Space, Thermal Phase Transition, And Confinement In Gauge Theories

Edward Witten[1]

School of Natural Sciences, Institute for Advanced Study
Olden Lane, Princeton, NJ 08540, USA

The correspondence between supergravity (and string theory) on AdS space and boundary conformal field theory relates the thermodynamics of $\mathcal{N} = 4$ super Yang-Mills theory in four dimensions to the thermodynamics of Schwarzschild black holes in Anti-de Sitter space. In this description, quantum phenomena such as the spontaneous breaking of the center of the gauge group, magnetic confinement, and the mass gap are coded in classical geometry. The correspondence makes it manifest that the entropy of a very large AdS Schwarzschild black hole must scale "holographically" with the volume of its horizon. By similar methods, one can also make a speculative proposal for the description of large N gauge theories in four dimensions without supersymmetry.

[1] Originally published in Adv. Theor. Math. Phys. **2** (1998) 505. Permission for use granted by International Press.

1. Introduction

Understanding the large N behavior of gauge theories in four dimensions is a classic and important problem [1]. The structure of the "planar diagrams" that dominate the large N limit gave the first clue that this problem might be solved by interpreting four-dimensional large N gauge theory as a string theory. Attempts in this direction have led to many insights relevant to critical string theory; for an account of the status, see [2].

Recently, motivated by studies of interactions of branes with external probes [3-7], and near-extremal brane geometry [8,10], a concrete proposal in this vein has been made [11], in the context of certain conformally-invariant theories such as $\mathcal{N} = 4$ super Yang-Mills theory in four dimensions. The proposal relates supergravity on anti-de Sitter or AdS space (or actually on AdS times a compact manifold) to conformal field theory on the boundary, and thus potentially introduces into the study of conformal field theory the whole vast subject of AdS compactification of supergravity (for a classic review see [12]; see also [13] for an extensive list of references relevant to current developments). Possible relations of a theory on AdS space to a theory on the boundary have been explored for a long time, both in the abstract (for example, see [14]), and in the context of supergravity and brane theory (for example, see [15]).

More complete references relevant to current developments can be found in papers already cited and in many of the other important recent papers [16-48] in which many aspects of the CFT/AdS correspondence have been extended and better understood.

In [29,49], a precise recipe was presented for computing CFT observables in terms of AdS space. It will be used in the present paper to study in detail a certain problem in gauge theory dynamics. The problem in question, already discussed in part in section 3.2 of [49], is to understand the high temperature behavior of $\mathcal{N} = 4$ super Yang-Mills theory. As we will see, in this theory, the CFT/AdS correspondence implies, in the infinite volume limit, many expected but subtle quantum properties, including a non-zero expectation value for a temporal Wilson loop [50,51], an area law for spatial Wilson loops, and a mass gap. (The study of Wilson loops is based on a formalism that was introduced recently [39,40].) These expectations are perhaps more familiar for ordinary four-dimensional Yang-Mills theory without supersymmetry – for a review see [52]. But the incorporation of supersymmetry, even $\mathcal{N} = 4$ supersymmetry, is not expected to affect these particular issues, since non-zero temperature breaks supersymmetry explicitly and makes it possible for the spin zero and spin one-half fields to get mass,[2] very plausibly reducing the high temperature behavior to that of the pure gauge theory. The ability to recover from the CFT/AdS correspondence relatively subtle dynamical properties of high temperature gauge theories, in a situation not governed by supersymmetry or conformal invariance, certainly illustrates the power of this correspondence.

In section 2, we review the relevant questions about gauge theories and the framework in which we will work, and develop a few necessary properties of the Schwarzschild black hole on AdS space. The CFT/AdS correspondence implies readily that in the limit of large mass, a Schwarzschild black hole in AdS space has an entropy proportional to the volume of the horizon, in agreement with the classic result of Bekenstein [57] and Hawking [58]. (The comparison of horizon volume of the AdS Schwarzschild solution to field theory entropy was first made, in the AdS_5 case, in [3], using a somewhat different language. As in some other string-theoretic studies of Schwarzschild black holes [53,54], and some earlier studies of BPS-saturated black holes [55], but unlike some microscopic studies of BPS black holes [56], in our discussion we are not able to determine the constant of proportionality between area and entropy.) This way of understanding black hole entropy is in keeping with the notion of "holography" [59-61]. The result holds for black holes with Schwarzschild radius much greater than the radius of curvature of the AdS space, and so does not immediately imply the corresponding result for Schwarzschild black holes in Minkowski space.

In section 3, we demonstrate, on the basis of the CFT/AdS correspondence, that the $\mathcal{N}=4$ theory at nonzero temperature has the claimed properties, especially the breaking of the center of the gauge group, magnetic confinement, and the mass gap.

In section 4, we present, using similar ideas, a proposal for studying ordinary large N gauge theory in four dimensions (without supersymmetry or matter fields) via string theory. In this proposal, we can exhibit confinement and the mass gap, precisely by the same arguments used in section 3, along with the expected large N scaling, but we are not able to effectively compute hadron masses or show that the model is asymptotically free.

---

[2] The thermal ensemble on a spatial manifold $\mathbf{R}^3$ can be described by path integrals on $\mathbf{R}^3\times\mathbf{S}^1$, with a radius for the $\mathbf{S}^1$ equal to $\beta=T^{-1}$, with T the temperature. The fermions obey antiperiodic boundary conditions around the $\mathbf{S}^1$ direction, and so get masses of order $1/T$ at tree level. The spin zero bosons get mass at the one-loop level.

2. High Temperatures And *AdS* Black Holes

2.1. $\mathbf{R}^3$ *and* $\mathbf{S}^3$

We will study the $\mathcal{N}=4$ theory at finite temperature on a spatial manifold $\mathbf{S}^3$ or $\mathbf{R}^3$. $\mathbf{R}^3$ will be obtained by taking an infinite volume limit starting with $\mathbf{S}^3$.

To study the theory at finite temperature on $\mathbf{S}^3$, we must compute the partition function on $\mathbf{S}^3 \times \mathbf{S}^1$ – with supersymmetry-breaking boundary conditions in the $\mathbf{S}^1$ directions. We denote the circumferences of $\mathbf{S}^1$ and $\mathbf{S}^3$ as β and β', respectively. By conformal invariance, only the ratio β/β' matters. To study the finite temperature theory on $\mathbf{R}^3$, we take the large β' limit, reducing to $\mathbf{R}^3 \times \mathbf{S}^1$.

Once we are on $\mathbf{R}^3 \times \mathbf{S}^1$, with circumference β for $\mathbf{S}^1$, the value of β can be scaled out via conformal invariance. Thus, the $\mathcal{N}=4$ theory on $\mathbf{R}^3$ cannot have a phase transition at any nonzero temperature. Even if one breaks conformal invariance by formulating the theory on $\mathbf{S}^3$ with some circumference β', there can be no phase transition as a function of temperature, since theories with finitely many local fields have in general no phase transitions as a function of temperature.

However, in the large N limit, it is possible to have phase transitions even in finite volume [62]. In section 3.2 of [49], it was shown that the $\mathcal{N}=4$ theory on $\mathbf{S}^3 \times \mathbf{S}^1$ has in the large N limit a phase transition as a function of β/β'. The large β/β' phase has some properties in common with the usual large β (or small temperature) phase of confining gauge theories, while the small β/β' phase is analogous to a deconfining phase.

When we go to $\mathbf{R}^3 \times \mathbf{S}^1$ by taking $\beta' \to \infty$ for fixed β, we get $\beta/\beta' \to 0$. So the unique nonzero temperature phase of the $\mathcal{N}=4$ theory on $\mathbf{R}^3$ is on the high temperature side of the phase transition and should be compared to the deconfining phase of gauge theories. Making this comparison will be the primary goal of section 3. We will also make some remarks in section 3 comparing the low temperature phase on $\mathbf{S}^3 \times \mathbf{S}^1$ to the confining phase of ordinary gauge theories. Here one can make some suggestive observations, but the scope is limited because in the particular $\mathcal{N}=4$ gauge theory under investigation, the low temperature phase on $\mathbf{S}^3$ arises only in finite volume, while most of the deep questions of statistical mechanics and quantum dynamics refer to the infinite volume limit.

2.2. Review Of Gauge Theories

We will now review the relevant expectations concerning finite temperature gauge theories in four dimensions.

Deconfinement at high temperatures can be usefully described, in a certain sense, in terms of spontaneous breaking of the center of the gauge group (or more precisely of the subgroup of the center under which all fields transform trivially). For our purposes, the gauge group will be $G = SU(N)$, and the center is $\Gamma = \mathbf{Z}_N$; it acts trivially on all fields, making possible the following standard construction.

Consider $SU(N)$ gauge theory on $Y \times \mathbf{S}^1$, with Y any spatial manifold. A conventional gauge transformation is specified by the choice of a map $g : Y \times \mathbf{S}^1 \to G$ which we write explicitly as $g(y, z)$, with y and z denoting respectively points in Y and in $\mathbf{S}^1$. (In describing a gauge transformation in this way, we are assuming that the G-bundle has been trivialized at least locally along Y; global properties along Y are irrelevant in the present discussion.) Such a map has $g(y, z + \beta) = g(y, z)$. However, as all fields transform trivially under the center of G, we can more generally consider gauge transformations by gauge functions $g(y, z)$ that obey

$$g(y, z + \beta) = g(y, z)h, \tag{2.1}$$

with h an arbitrary element of the center. Let us call the group of such extended gauge transformations (with arbitrary dependence on z and y and any h) $\overline{G}$ and the group of ordinary gauge transformations (with $h = 1$ but otherwise unrestricted) G'. The quotient $\overline{G}/G'$ is isomorphic to the center Γ of G, and we will denote it simply as Γ. Factoring out G' is natural because it acts trivially on all local observables and physical states (for physical states, G'-invariance is the statement of Gauss's law), while Γ can act nontrivially on such observables.

An order parameter for spontaneous breaking of Γ is the expectation value of a temporal Wilson line. Thus, let C be any oriented closed path of the form $y \times \mathbf{S}^1$ (with again y a fixed point in Y), and consider the operator

$$W(C) = \mathrm{Tr} P \exp \int_C A, \tag{2.2}$$

with A the gauge field and the trace taken in the N-dimensional fundamental representation of $SU(N)$. Consider a generalized gauge transformation of the form (2.1), with h an

N^{th} root of unity representing an element of the center of $SU(N)$. Action by such a gauge transformation multiplies the holonomy of A around C by h, so one has

$$W(C) \to hW(C). \tag{2.3}$$

Hence, the expectation value $\langle W(C)\rangle$ is an order parameter for the spontaneous breaking of the Γ symmetry.

Of course, such spontaneous symmetry breaking will not occur (for finite N) in finite volume. But a nonzero expectation value $\langle W(C)\rangle$ in the infinite volume limit, that is with Y replaced by $\mathbf{R}^3$, is an important order parameter for deconfinement. Including the Wilson line $W(C)$ in the system means including an external static quark (in the fundamental representation of $SU(N)$), so an expectation value for $W(C)$ means intuitively that the cost in free energy of perturbing the system by such an external charge is finite. In a confining phase, this free energy cost is infinite and $\langle W(C)\rangle = 0$. The $\mathcal{N} = 4$ theory on $\mathbf{R}^3$ corresponds to a high temperature or deconfining phase; we will confirm in section 3, using the CFT/AdS correspondence, that it has spontaneous breaking of the center.

Other important questions arise if we take the infinite volume limit, replacing X by $\mathbf{R}^3$. The theory at long distances along $\mathbf{R}^3$ is expected to behave like a pure $SU(N)$ gauge theory in three dimensions. At nonzero temperature, at least for weak coupling, the fermions get a mass at tree level from the thermal boundary conditions in the $\mathbf{S}^1$ direction, and the scalars (those present in four dimensions, as well as an extra scalar that arises from the component of the gauge field in the $\mathbf{S}^1$ direction) get a mass at one loop level; so the long distance dynamics is very plausibly that of three-dimensional gauge fields. The main expected features of three-dimensional pure Yang-Mills theory are confinement and a mass gap. The mass gap means simply that correlation functions $\langle \mathcal{O}(y,z)\mathcal{O}'(y',z)\rangle$ vanish exponentially for $|y-y'| \to \infty$. Confinement is expected to show up in an area law for the expectation value of a spatial Wilson loop. The area law means the following. Let C be now an oriented closed loop encircling an area A in $\mathbf{R}^3$, at a fixed point on $\mathbf{S}^1$. The area law means that if C is scaled up, keeping its shape fixed and increasing A, then the expectation value of $W(C)$ vanishes exponentially with A.

"Confinement" In Finite Volume

Finally, there is one more issue that we will address here. In the large N limit, a criterion for confinement is whether (after subtracting a constant from the ground state energy) the free energy is of order one – reflecting the contributions of color singlet hadrons

– or of order N^2 – reflecting the contributions of gluons. (This criterion has been discussed in [63].) In [49], it was shown that in the $\mathcal{N}=4$ theory on $\mathbf{S}^3 \times \mathbf{S}^1$, the large N theory has a low temperature phase with a free energy of order 1 – a "confining" phase – and a high temperature phase with a free energy of order N^2 – an "unconfining" phase.

Unconfinement at high temperatures comes as no surprise, of course, in this theory, and since the theory on $\mathbf{R}^3 \times \mathbf{S}^1$, at any temperature, is in the high temperature phase, we recover the expected result that the infinite volume theory is not confining. However, it seems strange that the finite volume theory on $\mathbf{S}^3$, at low temperatures, is "confining" according to this particular criterion.

This, however, is a general property of large N gauge theories on $\mathbf{S}^3$, at least for weak coupling (and presumably also for strong coupling). On a round three-sphere, the classical solution of lowest energy is unique up to gauge transformations (flat directions in the scalar potential are eliminated by the $R\phi^2$ coupling to scalars, R being the Ricci scalar), and is given by setting the gauge field A, fermions ψ, and scalars ϕ all to zero. This configuration is invariant under global $SU(N)$ gauge transformations. The Gauss law constraint in finite volume says that physical states must be invariant under the global $SU(N)$. There are no zero modes for any fields (for scalars this statement depends on the $R\phi^2$ coupling). Low-lying excitations are obtained by acting on the ground state with a finite number of A, ψ, and ϕ creation operators, and then imposing the constraint of global $SU(N)$ gauge invariance. The creation operators all transform in the adjoint representation, and so are represented in color space by matrices $M_1, M_2, \ldots, M_s$. $SU(N)$ invariants are constructed as traces, say $\mathrm{Tr}\, M_1 M_2 \ldots M_s$. The number of such traces is given by the number of ways to order the factors and is independent of N. So the multiplicity of low energy states is independent of N, as is therefore the low temperature free energy.

This result, in particular, is kinematic, and has nothing to do with confinement. To see confinement from the N dependence of the free energy, we must go to infinite volume. On $\mathbf{R}^3$, the Gauss law constraint does not say that the physical states are invariant under global $SU(N)$ transformations, but only that their global charge is related to the electric field at spatial infinity. If the free energy on $\mathbf{R}^3$ is of order 1 (and not proportional to N^2), this actually is an order parameter for confinement.

Now let us go back to finite volume and consider the behavior at high temperatures. At high temperatures, one cannot effectively compute the free energy by counting elementary excitations. It is more efficient to work in the "crossed channel." In $\mathbf{S}^3 \times \mathbf{S}^1$, with circumferences β' and β, if we take $\beta' \to \infty$ with fixed β, the free energy is proportional to the volume of $\mathbf{S}^3$ times the ground state energy density of the $2+1$-dimensional theory that is obtained by compactification on $\mathbf{S}^1$ (with circumference β and supersymmetry-breaking boundary conditions). That free energy is of order N^2 (the supersymmetry breaking spoils the cancellation between bosons and fermions already at the one-loop level, and the one-loop contribution is proportional to N^2). The volume of $\mathbf{S}^3$ is of order $(\beta')^3$. So the free energy on $\mathbf{S}^3 \times \mathbf{S}^1$ scales as $N^2(\beta')^3$ if one takes $\beta' \to \infty$ at fixed β, or in other words as $N^2\beta^{-3}$ if one takes $\beta \to 0$ at fixed β'. Presently we will recover this dependence on β by comparing to black holes.

2.3. AdS Correspondence And Schwarzshild Black Holes

The version of the CFT/AdS correspondence that we will use asserts that conformal field theory on an n-manifold M is to be studied by summing over contributions of Einstein manifolds B of dimension $n+1$ which (in a sense explained in [29,49]) have M at infinity.

We will be mainly interested in the case that $M = \mathbf{S}^{n-1} \times \mathbf{S}^1$, or $\mathbf{R}^{n-1} \times \mathbf{S}^1$. For $\mathbf{S}^{n-1} \times \mathbf{S}^1$, there are two known B's, identified by Hawking and Page [64] in the context of quantum gravity on AdS space. One manifold, X_1, is the quotient of AdS space by a subgroup of $SO(1, n+1)$ that is isomorphic to $\mathbf{Z}$. The metric (with Euclidean signature) can be written

$$ds^2 = \left(\frac{r^2}{b^2} + 1\right) dt^2 + \frac{dr^2}{\left(\frac{r^2}{b^2}\right) + 1} + r^2 d\Omega^2, \tag{2.4}$$

with $d\Omega^2$ the metric of a round sphere $\mathbf{S}^{n-1}$ of unit radius. Here t is a periodic variable of arbitrary period. We have normalized (2.4) so that the Einstein equations read

$$R_{ij} = -nb^{-2} g_{ij}; \tag{2.5}$$

here b is the radius of curvature of the anti-de Sitter space. With this choice, n does not appear explicitly in the metric. This manifold can contribute to either the standard thermal ensemble $\mathrm{Tr}e^{-\beta H}$ or to $\mathrm{Tr}(-1)^F e^{-\beta H}$, depending on the boundary conditions one uses for fermions in the t direction. The topology of X_1 is $\mathbf{R}^n \times \mathbf{S}^1$, or $\mathbf{B}^n \times \mathbf{S}^1$ ($\mathbf{B}^n$ denoting an n-ball) if we compactify it by including the boundary points at $r = \infty$.

The second solution, X_2, is the Schwarzschild black hole, in AdS space. The metric is

$$ds^2 = \left(\frac{r^2}{b^2} + 1 - \frac{w_n M}{r^{n-2}}\right) dt^2 + \frac{dr^2}{\left(\frac{r^2}{b^2} + 1 - \frac{w_n M}{r^{n-2}}\right)} + r^2 d\Omega^2. \tag{2.6}$$

Here w_n is the constant

$$w_n = \frac{16\pi G_N}{(n-1)\mathrm{Vol}(\mathbf{S}^{n-1})}. \tag{2.7}$$

Here G_N is the $n+1$-dimensional Newton's constant and $\mathrm{Vol}(\mathbf{S}^{n-1})$ is the volume of a unit $n-1$-sphere; the factor w_n is included so that M is the mass of the black hole (as we will compute later). Also, the spacetime is restricted to the region $r \geq r_+$, with r_+ the largest solution of the equation

$$\frac{r^2}{b^2} + 1 - \frac{w_n M}{r^{n-2}} = 0. \tag{2.8}$$

The metric (2.6) is smooth and complete if and only if the period of t is

$$\beta_0 = \frac{4\pi b^2 r_+}{n r_+^2 + (n-2)b^2}. \tag{2.9}$$

For future use, note that in the limit of large M one has

$$\beta_0 \sim \frac{4\pi b^2}{n(w_n b^2)^{1/n} M^{1/n}}. \tag{2.10}$$

As in the $n = 3$ case considered in [64], β_0 has a maximum as a function of r_+, so the Schwarzschild black hole only contributes to the thermodynamics if β is small enough, that is if the temperature is high enough. Moreover, X_2 makes the dominant contribution at sufficiently high temperature, while X_1 dominates at low temperature. The topology of X_2 is $\mathbf{R}^2 \times \mathbf{S}^{n-1}$, or $\mathbf{B}^2 \times \mathbf{S}^{n-1}$ if we compactify it to include boundary points. In particular, X_2 is simply-connected, has a unique spin structure, and contributes to the standard thermal ensemble but not to $\mathrm{Tr}(-1)^F e^{-\beta H}$.

With either (2.4) or (2.6), the geometry of the $\mathbf{S}^{n-1} \times \mathbf{S}^1$ factor at large r can be simply explained: the $\mathbf{S}^1$ has radius approximately $\beta = (r/b)\beta_0$, and the $\mathbf{S}^{n-1}$ has radius $\beta' = r/b$. The ratio is thus $\beta/\beta' = \beta_0$. If we wish to go to $\mathbf{S}^1 \times \mathbf{R}^{n-1}$, we must take $\beta/\beta' \to 0$, that is $\beta_0 \to 0$; this is the limit of large temperatures. (2.9) seems to show that this can be done with either $r_+ \to 0$ or $r_+ \to \infty$, but the $r_+ \to 0$ branch is thermodynamically unfavored [64] (having larger action), so we must take the large r_+ branch, corresponding to large M.

A scaling that reduces (2.9) to a solution with boundary $\mathbf{R}^{n-1} \times \mathbf{S}^1$ may be made as follows. If we set $r = (w_n M/b^{n-2})^{1/n}\rho$, $t = (w_n M/b^{n-2})^{-1/n}\tau$, then for large M we can reduce $r^2/b^2 + 1 - w_n M/r^{n-2}$ to $(w_n M/b^{n-2})^{2/n}(\rho^2/b^2 - b^{n-2}/\rho^{n-2})$. The period of τ become $\beta_1 = (w_n M/b^{n-2})^{1/n}\beta_0$ or (from (2.10)) for large M

$$\beta_1 = \frac{4\pi b}{n}. \tag{2.11}$$

The metric becomes

$$ds^2 = \left(\frac{\rho^2}{b^2} - \frac{b^{n-2}}{\rho^{n-2}}\right) d\tau^2 + \frac{d\rho^2}{\frac{\rho^2}{b^2} - \frac{b^{n-2}}{\rho^{n-2}}} + (w_n M/b^{n-2})^{2/n} \rho^2\, d\Omega^2. \tag{2.12}$$

The $M^{2/n}$ multiplying the last term means that the radius of $\mathbf{S}^{n-1}$ is of order $M^{1/n}$ and so diverges for $M \to \infty$. Hence, the $\mathbf{S}^{n-1}$ is becoming flat and looks for $M \to \infty$ locally like $\mathbf{R}^{n-1}$. If we introduce near a point $P \in \mathbf{S}^{n-1}$ coordinates y_i such that at P, $d\Omega^2 = \sum_i dy_i^2$, and then set $y_i = (w_n M/b^{n-2})^{-1/n} x_i$, then the metric becomes

$$ds^2 = \left(\frac{\rho^2}{b^2} - \frac{b^{n-2}}{\rho^{n-2}}\right) d\tau^2 + \frac{d\rho^2}{\left(\frac{\rho^2}{b^2} - \frac{b^{n-2}}{\rho^{n-2}}\right)} + \rho^2 \sum_{i=1}^{n-1} dx_i^2. \tag{2.13}$$

This is the desired solution $\widetilde{X}$ that is asymptotic at infinity to $\mathbf{R}^{n-1} \times \mathbf{S}^1$ instead of $\mathbf{S}^{n-1} \times \mathbf{S}^1$. Its topology, if we include boundary points, is $\mathbf{R}^{n-1} \times \mathbf{B}^2$. The same solution was found recently by scaling of a near-extremal brane solution [45].

2.4. Entropy Of Schwarzschild Black Holes

Following Hawking and Page [64] (who considered the case $n = 3$), we will now describe the thermodynamics of Schwarzschild black holes in AdS_{n+1}. Our normalization of the cosmological constant is stated in (2.5). The bulk Einstein action with this value of the cosmological constant is

$$I = -\frac{1}{16\pi G_N} \int d^{n+1}x \sqrt{g} \left(R + \frac{\frac{1}{2}n(n-1)}{b^2}\right). \tag{2.14}$$

For a solution of the equations of motion, one has $R = -\frac{1}{2}n(n+1)/b^2$, and the action becomes

$$I = \frac{n}{8\pi G_N} \int d^{n+1}x \sqrt{g}, \tag{2.15}$$

that is, the volume of spacetime times $n/8\pi G_N$. The action additionally has a surface term [65,66], but the surface term vanishes for the AdS Schwarzschild black hole, as noted in [64], because the black hole correction to the AdS metric vanishes too rapidly at infinity.

Actually, both the AdS spacetime (2.4) and the black hole spacetime (2.6) have infinite volume. As in [64], one subtracts the two volumes to get a finite result. Putting an upper cutoff R on the radial integrations, the regularized volume of the AdS spacetime is

$$V_1(R) = \int_0^{\beta'} dt \int_0^R dr \int_{\mathbf{S}^{n-1}} d\Omega \, r^{n-1}, \tag{2.16}$$

and the regularized volume of the black hole spacetime is

$$V_2(R) = \int_0^{\beta_0} dt \int_{r_+}^R dr \int_{\mathbf{S}^{n-1}} d\Omega \, r^{n-1}. \tag{2.17}$$

One difference between the two integrals is obvious here: in the black hole spacetime $r \geq r_+$, while in the AdS spacetime $r \geq 0$. A second and slightly more subtle difference is that one must use different periodicities β' and β_0 for the t integrals in the two cases. The black hole spacetime is smooth only if β_0 has the value given in (2.9), but for the AdS spacetime, any value of β' is possible. One must adjust β' so that the geometry of the hypersurface $r = R$ is the same in the two cases; this is done by setting $\beta' \sqrt{(r^2/b^2) + 1} = \beta_0 \sqrt{(r^2/b^2) + 1 - w_n M / r^{n-2}}$. After doing so, one finds that the action difference is

$$I = \frac{n}{8\pi G_N} \lim_{R \to \infty} (V_2(R) - V_1(R)) = \frac{\mathrm{Vol}(\mathbf{S}^{n-1})(b^2 r_+^{n-1} - r_+^{n+1})}{4G_N(nr_+^2 + (n-2)b^2)}. \tag{2.18}$$

This is positive for small r_+ and negative for large r_+, showing that the phase transition found in [64] occurs for all n.

Then, as in [64], one computes the energy

$$E = \frac{\partial I}{\partial \beta_0} = \frac{(n-1)\mathrm{Vol}(\mathbf{S}^{n-1})(r_+^n b^{-2} + r_+^{n-2})}{16\pi G_N} = M \tag{2.19}$$

and the entropy

$$S = \beta_0 E - I = \frac{1}{4G_N} r_+^{n-1} \mathrm{Vol}(\mathbf{S}^{n-1}) \tag{2.20}$$

of the black hole. The entropy can be written

$$S = \frac{A}{4G_N}, \tag{2.21}$$

with A the volume of the horizon, which is the surface at $r = r_+$.

Comparison To Conformal Field Theory

Now we can compare this result for the entropy to the predictions of conformal field theory.

The black hole entropy should be compared to boundary conformal field theory on $\mathbf{S}^{n-1} \times \mathbf{S}^1$, where the two factors have circumference 1 and β_0/b, respectively. In the limit as $\beta_0 \to 0$, this can be regarded as a high temperature system on $\mathbf{S}^{n-1}$. Conformal invariance implies that the entropy density on $\mathbf{S}^{n-1}$ scales, in the limit of small β_0, as $\beta_0^{-(n-1)}$. According to (2.9), $\beta_0 \to 0$ means $r_+ \to \infty$ with $\beta_0 \sim 1/r_+$. Hence, the boundary conformal field theory predicts that the entropy of this system is of order r_+^{n-1}, and thus asymptotically is a fixed multiple of the horizon volume which appears in (2.21). This is of course the classic result of Bekenstein and Hawking, for which microscopic explanations have begun to appear only recently. Note that this discussion assumes that $\beta_0 << 1$, which means that $r_+ >> b$; so it applies only to black holes whose Schwarzschild radius is much greater than the radius of curvature of AdS space. However, in this limit, one does get a simple explanation of why the black hole entropy is proportional to area. The explanation is entirely "holographic" in spirit [59,61].

To fix the constant of proportionality between entropy and horizon volume (even in the limit of large black holes), one needs some additional general insight, or some knowledge of the quantum field theory on the boundary. For 2+1-dimensional black holes, in the context of an old framework [69] for a relation to boundary conformal field theory which actually is a special case of the general CFT/AdS correspondence, such additional information is provided by modular invariance of the boundary conformal field theory [67,68].

3. High Temperature Behavior Of The $\mathcal{N} = 4$ Theory

In this section we will address three questions about the high temperature behavior of the $\mathcal{N} = 4$ theory that were raised in section 2: the behavior of temporal Wilson lines; the behavior of spatial Wilson lines; and the existence of a mass gap.

In discussing Wilson lines, we use a formalism proposed recently [39,40]. Suppose one is doing physics on a four-manifold M which is the boundary of a five-dimensional Einstein manifold B (of negative curvature). To compute a Wilson line associated with a contour $C \subset M$, we study elementary strings on B with the property that the string worldsheet D has C for its boundary. Such a D has an infinite area, but the divergence is proportional to the circumference of C. One can define therefore a regularized area $\alpha(D)$ by subtracting from the area of D an infinite multiple of the circumference of C. The expectation value of a Wilson loop $W(C)$ is then roughly

$$\langle W(C)\rangle = \int_{\mathcal{D}} d\mu\, e^{-\alpha(D)} \tag{3.1}$$

where $\mathcal{D}$ is the space of string worldsheets obeying the boundary conditions and $d\mu$ is the measure of the worldsheet path integral. Moreover, according to [39,40], in the regime in which supergravity is valid (large N and large g^2N), the integral can be evaluated approximately by setting D to the surface of smallest $\alpha(D)$ that obeys the boundary conditions.

The formula (3.1) is oversimplified for various reasons. For one thing, worldsheet fermions must be included in the path integral. Also, the description of the $\mathcal{N}=4$ theory actually involves not strings on B but strings on the ten-manifold $B \times \mathbf{S}^5$. Accordingly, what are considered in [39,40] are some generalized Wilson loop operators with scalar fields included in the definition; the boundary behavior of D in the $\mathbf{S}^5$ factor depends on which operator one uses. But if all scalars have masses, as they do in the $\mathcal{N}=4$ theory at positive temperature, the generalized Wilson loop operators are equivalent at long distances to conventional ones. An important conclusion from (3.1) nonetheless stands: Wilson loops on $\mathbf{R}^4$ will obey an area law if, when C is scaled up, the minimum value of $\alpha(D)$ scales like a positive multiple of the area enclosed by C. (3.1) also implies vanishing of $\langle W(C)\rangle$ if suitable D's do not exist, that is, if C is not a boundary in B.

3.1. Temporal Wilson Lines

Our first goal will be to analyze temporal Wilson lines. That is, we take spacetime to be $\mathbf{S}^3 \times \mathbf{S}^1$ or $\mathbf{R}^3 \times \mathbf{S}^1$, and we take $C = P \times \mathbf{S}^1$, with P a point in $\mathbf{S}^3$ or in $\mathbf{R}^3$.

We begin on $\mathbf{S}^3$ in the low temperature phase. We recall that this is governed by a manifold X_1 with the topology of $\mathbf{B}^4 \times \mathbf{S}^1$. In particular, the contour C, which wraps around the $\mathbf{S}^1$, is not homotopic to zero in X_1 and is not the boundary of any D. Thus, the expectation value of a temporal Wilson line vanishes at low temperatures. This is the expected result, corresponding to the fact that the center Γ of the gauge group is unbroken at low temperatures.

Now we move on to the high temperature phase on $\mathbf{S}^3$. This phase is governed by a manifold X_2 that is topologically $\mathbf{S}^3 \times \mathbf{B}^2$. In this phase, $C = P \times \mathbf{S}^1$ is a boundary; in fact it is the boundary of $D = P \times \mathbf{B}^2$. Thus, it appears at first sight that the temporal

Wilson line has a vacuum expectation value and that the center of the gauge group is spontaneously broken.

There is a problem here. Though we expect these results in the high temperature phase on $\mathbf{R}^3$, they cannot hold on $\mathbf{S}^3$, because the center (or any other bosonic symmetry) cannot be spontaneously broken in finite volume. The resolution of the puzzle is instructive. The classical solution on X_2 is not unique. We must recall that Type IIB superstring theory has a two-form field B that couples to the elementary string world-sheet D by

$$i\int_D B. \tag{3.2}$$

The gauge-invariant field strength is $H = dB$. We can add to the solution a "world-sheet theta angle," that is a B field of $H = 0$ with an arbitrary value of $\psi = \int_D B$ (here D is any surface obeying the boundary conditions, for instance $D = P \times \mathbf{B}^2$). Since discrete gauge transformations that shift the flux of B by a multiple of 2π are present in the theory, ψ is an angular variable with period 2π.

If this term is included, the path integrand in (3.1) receives an extra factor $e^{i\psi}$. Upon integrating over the space of all classical solutions – that is integrating over the value of ψ – the expectation value of the temporal Wilson line on $\mathbf{S}^3$ vanishes.

Now, let us go to $\mathbf{R}^3 \times \mathbf{S}^1$, which is the boundary of $\mathbf{R}^3 \times \mathbf{B}^2$. In infinite volume, ψ is best understood as a massless scalar field in the low energy theory on $\mathbf{R}^3$. One still integrates over local fluctuations in ψ, but not over the vacuum expectation value of ψ, which is set by the value at spatial infinity. The expectation value of $W(C)$ is nonzero and is proportional to $e^{i\psi}$.

What we have seen is thus spontaneous symmetry breaking: in infinite volume, the expectation value of $\langle W(C)\rangle$ is nonzero, and depends on the choice of vacuum, that is on the value of ψ. The field theory analysis that we reviewed in section 2 indicates that the symmetry that is spontaneously broken by the choice of ψ is the center, Γ, of the gauge group. Since ψ is a continuous angular variable, it seems that the center is $U(1)$. This seems to imply that the gauge group is not $SU(N)$, with center $\mathbf{Z}_N$, but $U(N)$. However, a variety of arguments [49] show that the AdS theory encodes a $SU(N)$ gauge group, not $U(N)$. Perhaps the apparent $U(1)$ center should be understood as a large N limit of $\mathbf{Z}_N$.

't Hooft Loops

We would also like to consider in a similar way 't Hooft loops. These are obtained from Wilson loops by electric-magnetic duality. Electric-magnetic duality of $\mathcal{N} = 4$ arises

[70,71] directly from the $\tau \to -1/\tau$ symmetry of Type IIB. That symmetry exchanges elementary strings with D-strings. So to study the 't Hooft loops we need only, as in [48], replace elementary strings by D-strings in the above discussion.

The $\tau \to -1/\tau$ symmetry exchanges the Neveu-Schwarz two-form B which entered the above discussion with its Ramond-Ramond counterpart B'; the D-brane theta angle $\psi' = \int_D B'$ thus plays the role of ψ in the previous discussion. In the thermal physics on $\mathbf{R}^3 \times \mathbf{S}^1$, the center of the "magnetic gauge group" is spontaneously broken, and the temporal 't Hooft loops have an expectation value, just as we described for Wilson loops. The remarks that we make presently about spatial Wilson loops similarly carry over for spatial 't Hooft loops.

3.2. Spatial Wilson Loops

Now we will investigate the question of whether at nonzero temperature the spatial Wilson loops obey an area law. The main point is to first understand why there is *not* an area law at zero temperature. At zero temperature, one works with the AdS metric

$$ds^2 = \frac{1}{x_0^2}\left(dx_0^2 + \sum_{i=1}^{4} dx_i^2\right). \tag{3.3}$$

We identify the spacetime M of the $\mathcal{N} = 4$ theory with the boundary at $x_0 = 0$, parametrized by the Euclidean coordinates x_i, $i = 1, \ldots, 4$. M has a metric $d\widetilde{s}^2 = \sum_i dx_i^2$ obtained by multiplying ds^2 by x_0^2 and setting $x_0 = 0$. (If we use a function other than x_0^2, the metric on M changes by a conformal transformation.) We take a closed oriented curve $C \subset M$ and regard it as the boundary of an oriented compact surface D in AdS space. The area of D is infinite, but after subtracting an infinite counterterm proportional to the circumference of C, we get a regularized area $\alpha(D)$. In the framework of [39,40], the expectation value of the Wilson line $W(C)$ is proportional to $\exp(-\alpha(D))$, with D chosen to minimize $\alpha(D)$.

Now the question arises: why does not this formalism *always* give an area law? As the area enclosed by C on the boundary is scaled up, why is not the area of D scaled up proportionately? The answer to this is clear from conformal invariance. If we scale up C via $x_i \to tx_i$, with large positive t, then by conformal invariance we can scale up D, with $x_i \to tx_i$, $x_0 \to tx_0$, without changing its area (except for a boundary term involving the regularization). Thus the area of D need not be proportional to the area enclosed by C on the boundary. Since, however, in this process we had to scale $x_0 \to tx_0$ with very large t, the surface D which is bounded by a very large circle C "bends" very far away from the boundary of AdS space. If such a bending of D were prevented – if D were limited to a region with $x_0 \leq L$ for some cutoff L – then one would get an area law for $W(C)$. This is precisely what will happen at nonzero temperature.

At nonzero temperature, we have in fact the metric (2.13) obtained earlier, with $n = 4$:

$$ds^2 = \left(\frac{\rho^2}{b^2} - \frac{b^2}{\rho^2}\right) d\tau^2 + \frac{d\rho^2}{\left(\frac{\rho^2}{b^2} - \frac{b^2}{\rho^2}\right)} + \rho^2 \sum_{i=1}^{3} dx_i^2. \tag{3.4}$$

The range of ρ is $b \leq \rho \leq \infty$. Spacetime – a copy of $\mathbf{R}^3 \times \mathbf{S}^1$ – is the boundary at $\rho = \infty$. We define a metric on spacetime by dividing by ρ^2 and setting $\rho = \infty$. In this way we obtain the spacetime metric

$$d\widetilde{s}^2 = \frac{d\tau^2}{b^2} + \sum_{i=1}^{3} dx_i^2. \tag{3.5}$$

As the period of τ is β_1, the circumference of the $\mathbf{S}^1$ factor in $\mathbf{R}^3 \times \mathbf{S}^1$ is β_1/b and the temperature is

$$T = \frac{b}{\beta_1} = \frac{1}{\pi}. \tag{3.6}$$

Because of conformal invariance, the numerical value of course does not matter.

Now, let C be a Wilson loop in $\mathbf{R}^3$, at a fixed value of τ, enclosing an area A in $\mathbf{R}^3$. A bounding surface D in the spacetime (3.4) is limited to $\rho \geq b$, so the coefficient of $\sum_i dx_i^2$ is always at least b^2. Apart from a surface term that depends on the regularization and the detailed solution of the equation for a minimal surface, the regularized area of D is at least $\alpha(D) = b^2 A$ (and need be no larger than this). The Wilson loops therefore obey an area law, with string tension b^2 times the elementary Type IIB string tension.

We could of course have used a function other than ρ^2 in defining the spacetime metric, giving a conformally equivalent metric on spacetime. For instance, picking a constant s and using $s^2\rho^2$ instead of ρ^2 would scale the temperature as $T \to T/s$ and would multiply all lengths on $\mathbf{R}^3$ by s. The area enclosed by C would thus become $A' = As^2$. As $\alpha(D)$ is unaffected, the relation betseen $\alpha(D)$ and A' becomes $\alpha(D) = \left(b^2/s^2\right) \cdot A'$. The string tension in the Wilson loop area law thus scales like s^{-2}, that is, like T^2, as expected from conformal invariance.

3.3. The Mass Gap

The last issue concerning the $\mathcal{N} = 4$ theory at high temperature that we will discuss here is the question of whether there is a mass gap. We could do this by analyzing correlation functions, using the formulation of [29,49], but it is more direct to use a Hamiltonian approach (discussed at the end of [49]) in which one identifies the quantum states of the supergravity theory with those of the quantum field theory on the boundary.

So we will demonstrate a mass gap by showing that there is a gap, in the three-dimensional sense, for quantum fields propagating on the five-dimensional spacetime

$$ds^2 = \left(\frac{\rho^2}{b^2} - \frac{b^2}{\rho^2}\right) d\tau^2 + \frac{d\rho^2}{\left(\frac{\rho^2}{b^2} - \frac{b^2}{\rho^2}\right)} + \rho^2 \sum_{i=1}^{3} dx_i^2. \tag{3.7}$$

This spacetime is the product of a three-space $\mathbf{R}^3$, parametrized by the x_i, with a two-dimensional "internal space" $\mathbf{W}$, parametrized by ρ and τ. We want to show that a quantum free field propagating on this five-dimensional spacetime gives rise, in the three-dimensional sense, to a discrete spectrum of particle masses, all of which are positive. When such a spectrum is perturbed by interactions, the discreteness of the spectrum is lost (as the very massive particles become unstable), but the mass gap persists.

If $\mathbf{W}$ were compact, then discreteness of the mass spectrum would be clear: particle masses on $\mathbf{R}^3$ would arise from eigenvalues of the Laplacian (and other wave operators) on $\mathbf{W}$. Since $\mathbf{W}$ is not compact, it is at first sight surprising that a discrete mass spectrum will emerge. However, this does occur, by essentially the same mechanism that leads to discreteness of particle energy levels on *AdS* space [72,73] with a certain notion of energy.

For illustrative purposes, we will consider the propagation of a Type IIB dilaton field ϕ on this spacetime. Other cases are similar. The action for ϕ is

$$\begin{aligned} I(\phi) =& \frac{1}{2}\int_b^\infty d\rho \int_0^{\beta_1/b} d\tau \int_{-\infty}^{\infty} d^3x \; \rho^3 \left(\left(\frac{\rho^2}{b^2} - \frac{b^2}{\rho^2}\right)\left(\frac{\partial\phi}{\partial\rho}\right)^2 \right. \\ & \left. + \left(\frac{\rho^2}{b^2} - \frac{b^2}{\rho^2}\right)^{-1}\left(\frac{\partial\phi}{\partial\tau}\right)^2 + \rho^{-2}\sum_i \left(\frac{\partial\phi}{\partial x_i}\right)^2 \right). \end{aligned} \tag{3.8}$$

Since translation of τ is a symmetry, modes with different momentum in the τ direction are decoupled from one another. The spectrum of such momenta is discrete (as τ is a periodic variable). To simplify things slightly and illustrate the essential point, we will write the formulas for the modes that are independent of τ; others simply give, by the same argument, additional three-dimensional massive particles with larger masses.

We look for a solution of the form $\phi(\rho, x) = f(\rho)e^{ik\cdot x}$, with $\vec{k}$ the momentum in $\mathbf{R}^3$. The effective Lagrangian becomes

$$I(f) = \frac{1}{2}\int_b^\infty d\rho\, \rho^3 \left((\rho^2/b^2 - b^2/\rho^2)\left(\frac{df}{d\rho}\right)^2 + \rho^{-2}k^2f^2\right). \tag{3.9}$$

The equation of motion for f is

$$-\rho^{-1}\frac{d}{d\rho}\left(\rho^3(\rho^2/b^2 - b^2/\rho^2)\frac{df}{d\rho}\right) + k^2 f = 0. \tag{3.10}$$

A mode of momentum k has a mass m, in the three-dimensional sense, that is given by $m^2 = -k^2$. We want to show that the equation (3.10) has acceptable solutions only if m^2 is in a certain discrete set of positive numbers.

Acceptable solutions are those that obey the following boundary conditions:

(1) At the lower endpoint $\rho = b$, we require $df/d\rho = 0$. The reason for this is that ρ behaves near this endpoint as the origin in polar coordinates; hence f is not smooth at this endpoint unless $df/d\rho = 0$ there.

(2) For $\rho \to \infty$, the equation has two linearly independent solutions, which behave as $f \sim$ constant and $f \sim \rho^{-4}$. We want a normalizable solution, so we require that $f \sim \rho^{-4}$.

For given k^2, the equation (3.10) has, up to a constant multiple, a unique solution that obeys the correct boundary condition near the lower endpoint. For generic k^2, this solution will approach a nonzero constant for $\rho \to \infty$. As in standard quantum mechanical problems, there is a normalizable solution only if k^2 is such that the solution that behaves correctly at the lower endpoint also vanishes for $\rho \to \infty$. This "eigenvalue" condition determines a discrete set of values of k^2.

The spectrum thus consists entirely of a discrete set of normalizable solutions. There are no such normalizable solutions for $k^2 \geq 0$. This can be proved by noting that, given a normalizable solution f of the equation of motion, a simple integration by parts shows that the action (3.9) vanishes. For $k^2 \geq 0$, vanishing of $I(f)$ implies that $df/d\rho = 0$, whence (given normalizability) $f = 0$. So the discrete set of values of k^2 at which there are normalizable solutions are all negative; the masses $m^2 = -k^2$ are hence strictly positive. This confirms the existence of the mass gap.

To understand the phenomenon better, let us compare to what usually happens in quantum mechanics. In typical quantum mechanical scattering problems, with potentials that vanish at infinity, the solutions with positive energy (analogous to $m^2 > 0$) are oscillatory at infinity and obey plane wave normalizability. When this is so, both solutions at infinity are physically acceptable (in some situations, for example, they are interpreted as incoming and outgoing waves), and one gets a continuous spectrum that starts at zero energy. The special property of the problem we have just examined is that even for negative k^2, there are no oscillatory solutions at infinity, and instead one of the two solutions must be rejected as being unnormalizable near $\rho = \infty$. This feature leads to the discrete spectrum.

If instead of the spacetime (3.7), we work on AdS spacetime (2.4), there is a continuous spectrum of solutions with plane wave normalizability for all $k^2 < 0$; this happens because for $k^2 < 0$ one gets oscillatory solutions near the lower endpoint, which for the AdS case is at $r = 0$. Like confinement, the mass gap of the thermal $\mathcal{N} = 4$ theory depends on the cutoff at small r.

4. Approach To QCD

One interesting way to study four-dimensional gauge theory is by compactification from a certain exotic six-dimensional theory with $(0, 2)$ supersymmetry. This theory can be realized in Type IIB compactification on K3 [74] or in the dynamics of parallel M-theory fivebranes [75] and can apparently be interpreted [11] in terms of M-theory on $AdS_7 \times \mathbf{S}^4$. This interpretation is effective in the large N limit – as the M-theory radius of curvature is of order $N^{1/3}$. Since compactification from six to four dimensions has been an effective approach to gauge theory dynamics (for instance, in deducing Montonen-Olive duality [74] using a strategy proposed in [76]), it is natural to think of using the solution for the large N limit of the six-dimensional theory as a starting point to understand the four-dimensional theory.

Our basic approach will be as follows. If we compactify the six-dimensional $(0, 2)$ theory on a circle C_1 of radius R_1, with a supersymmetry-preserving spin structure (fermions are periodic in going around the circle), we get a theory that at low energies looks like five-dimensional $SU(N)$ supersymmetric Yang-Mills theory, with maximal supersymmetry and five-dimensional gauge coupling constant $g_5^2 = R_1$. Now compactify on a second circle C_2, orthogonal to the first, with radius R_2. If we take $R_2 >> R_1$, we can determine what the resulting four-dimensional theory is in a two-step process, compactifying to five dimensions on C_1 to get five-dimensional supersymmetric Yang-Mills theory and then compactifying to four-dimensions on C_2.

No matter what spin structure we use on C_2, we will get a four-dimensional $SU(N)$ gauge theory with gauge coupling $g_4^2 = R_1/R_2$. If we take on C_2 (and more precisely, on $C_1 \times C_2$) the supersymmetry-preserving spin structure, then the low energy theory will be the four-dimensional $\mathcal{N} = 4$ theory some of whose properties we have examined in the present paper. We wish instead to break supersymmetry by taking the fermions to be antiperiodic in going around C_2. Then the fermions get masses (of order $1/R_2$) at tree level, and the spin zero bosons very plausibly get masses (of order g_4^2N/R_2) at one-loop level. If this is so, the low energy theory will be the pure $SU(N)$ theory without supersymmetry. If $g_4^2 << 1$, the theory will flow at very long distances to strong coupling; at such long distances the spin one-half and spin one fields that receive tree level or one-loop masses will be irrelevant. So this is a possible framework for studying the pure Yang-Mills theory without supersymmetry.

We want to take the large N limit with $g_4 \to 0$ in such a way that $\eta = g_4^2N$ has a limit. So we need $g_4^2 = \eta/N$, or in other words

$$R_1 = \frac{\eta R_2}{N}. \tag{4.1}$$

We actually want η fixed and small, so that the four-dimensional Yang-Mills theory is weakly coupled at the compactification scale, and flows to strong coupling only at very long distances at which the detailed six-dimensional setup is irrelevant.

To implement this approach, we first look for an Einstein manifold that is asymptotic at infinity to $\mathbf{R}^5 \times C_2$. Though it may seem to reverse the logic of the construction, starting with C_2 first in constructing the solution turns out to be more convenient. The supersymmetry-breaking boundary conditions on C_2 are the right ones for using the spacetime (2.13) that is constructed by scaling of the seven-dimensional AdS Schwarzschild solution:

$$ds^2 = \left(\frac{\rho^2}{b^2} - \frac{b^4}{\rho^4}\right) d\tau^2 + \frac{d\rho^2}{\left(\frac{\rho^2}{b^2} - \frac{b^4}{\rho^4}\right)} + \rho^2 \sum_{i=1}^{5} dx_i^2. \tag{4.2}$$

According to [11], we want here

$$b = 2G_N^{1/9}(\pi N)^{1/3}. \tag{4.3}$$

(Here G_N is the eleven-dimensional Newton constant, so $G_N^{1/9}$ has dimensions of length. We henceforth set $G_N = 1$.)

To make the scaling with N clearer, we also set $\rho = 2(\pi N)^{1/3}\lambda$. And – noting from (2.11) that τ has period $4\pi b/n = (4/3)\pi^{4/3}N^{1/3}$ – we set

$$\tau = \theta \cdot \left(\frac{2\pi N}{3}\right)^{1/3}, \tag{4.4}$$

where θ is an ordinary angle, of period 2π. After also a rescaling of the x_i, the metric becomes

$$ds^2 = \frac{4}{9}\pi^{2/3}N^{2/3}\left(\lambda^2 - \frac{1}{\lambda^4}\right)d\theta^2 + 4\pi^{2/3}N^{2/3}\frac{d\lambda^2}{\left(\lambda^2 - \frac{1}{\lambda^4}\right)} + 4\pi^{2/3}N^{2/3}\lambda^2\sum_{i=1}^{5}dx_i^2. \tag{4.5}$$

Now we want to compactify one of the x_i, say x_5, on a second circle whose radius as measured at $\lambda = \infty$ should according to (4.1) should be η/N times the radius of the circle parametrized by θ. To do this, we write $x_5 = (\eta/N)\psi$ with ψ of period 2π. We also now restore the $\mathbf{S}^4$ factor that was present in the original M-theory on $AdS_7 \times \mathbf{S}^4$ and has so far been suppressed. The metric is now

$$\begin{aligned} ds^2 =& \frac{4}{9}\pi^{2/3}N^{2/3}\left(\lambda^2 - \frac{1}{\lambda^4}\right)d\theta^2 + \frac{4}{9}\eta^2\pi^{2/3}N^{-4/3}\lambda^2 d\psi^2 \\ &+ 4N^{2/3}\frac{d\lambda^2}{\left(\lambda^2 - \frac{1}{\lambda^4}\right)} + 4\pi^{2/3}N^{2/3}\rho^2\sum_{i=1}^{4}dx_i^2 + \pi^{2/3}N^{2/3}d\Omega_4^2. \end{aligned} \tag{4.6}$$

At this stage, θ and ψ are both ordinary angular variables of radius 2π, and $d\Omega_4^2$ is the metric on a unit four-sphere.

Now, we want to try to take the limit as $N \to \infty$. The metric becomes large in all directions except that one circle factor – the circle C_1, parametrized by ψ – shrinks. Thus we should try to use the equivalence between M-theory compactified on a small circle and weakly coupled Type IIA superstrings. We see that the radius $R(\lambda)$ of the circle parametrized by C is in fact

$$R(\lambda) = \frac{2}{3}\eta\lambda\pi^{1/3}N^{-2/3}. \tag{4.7}$$

To relate an M-theory compactification on a circle to a Type IIA compactification, we must [77] multiply the metric by R. All factors of N felicitously disappear from the metric, which becomes

$$ds^2 = \frac{8}{27}\eta\lambda\pi\left(\lambda^2 - \frac{1}{\lambda^4}\right)d\theta^2 + \frac{8\pi}{3}\eta\lambda\frac{d\lambda^2}{\left(\lambda^2 - \frac{1}{\lambda^4}\right)} + \frac{8\pi}{3}\eta\lambda^3\sum_{i=1}^{4}dx_i^2 + \frac{2\pi}{3}\eta\lambda d\Omega_4^2. \tag{4.8}$$

The string coupling constant is meanwhile

$$g_{st}^2 = R^{3/2} = \frac{(2/3)^{3/2}\eta^{3/2}\lambda^{3/2}\pi^{1/2}}{N}. \tag{4.9}$$

This result clearly has some of the suspected properties of large N gauge theories. The metric (4.8) is independent of N, so in the weak coupling limit, the spectrum of the string theory will be independent of N. Meanwhile, the string coupling constant (4.9) is of order $1/N$, as expected [1] for the residual interactions between color singlet states in the large N limit. The very ability to get a description such as this one in which $1/N$ only enters as a coupling constant (and not explicitly in the multiplicity of states) is a reflection of confinement. Confinement in the form of an area law for Wilson loops can be demonstrated along the lines of our discussion in section 3: it follows from the fact that the coefficient in the metric of $\sum_{i=1}^4 dx_i^2$ is bounded strictly above zero. A mass gap likewise can be demonstrated, as in section 3, by using the large λ behavior of the metric.

On the other hand, it is not obvious how one could hope to compute the spectrum or even show asymptotic freedom. Asymptotic freedom should say that as $\eta \to 0$, the particle masses become exponentially small (with an exponent determined by the gauge theory beta function). It is not at all clear how to demonstrate this. A clue comes from the fact that the coupling of the physical hadrons should be independent of η (and of order $1/N$) as $\eta \to 0$. In view of the formula (4.9), this means that we should take $\eta\lambda$ of order one as $\eta \to 0$. If we set $\widetilde{\lambda} = \eta\lambda$, and write the metric in terms of $\widetilde{\lambda}$, then the small η limit becomes somewhat clearer: a singularity develops at small $\widetilde{\lambda}$ for $\eta \to 0$. Apparently, in this approach, the mysteries of four-dimensional quantum gauge theory are encoded in the behavior of string theory near this singularity.

This singularity actually has a very simple and intuitive interpretation which makes it clearer why four-dimensional gauge theory can be described by string theory in the spacetime (4.8). The Euclidean signature Type IIA nonextremal fourbrane solution is described by the metric [78]

$$\begin{aligned} ds^2 = &\left(1 - \left(\frac{r_+}{r}\right)^3\right)\left(1 - \left(\frac{r_-}{r}\right)^3\right)^{-1/2} dt^2 + \left(1 - \left(\frac{r_+}{r}\right)^3\right)^{-1}\left(1 - \left(\frac{r_-}{r}\right)^3\right)^{-5/6} dr^2 \\ &+ \left(1 - \left(\frac{r_-}{r}\right)^3\right)^{1/2} \sum_{i=1}^4 dx_i^2 + r^2\left(1 - \left(\frac{r_-}{r}\right)^3\right)^{1/6} d\Omega_4^2, \end{aligned} \tag{4.10}$$

with $r_+ > r_- > 0$. The string coupling constant is

$$g_{st}^2 = \left(1 - \left(\frac{r_-}{r}\right)^3\right)^{1/2}. \tag{4.11}$$

The horizon is at $r = r_+$, and the spacetime is bounded by $r \geq r_+$. This spacetime is complete and smooth if t has period

$$T = 12\pi \left(1 - \left(\frac{r_-}{r_+}\right)^3\right)^{-1/6}. \tag{4.12}$$

If one continues (via Lorentzian or complex values of the coordinates) past $r = r_+$, there is a singularity at $r = r_-$. The extremal fourbrane solution is obtained by setting $r_+ = r_-$ and is singular. But this singularity is exactly the singularity that arises in (4.8) upon taking $\eta \to 0$, with $\eta\lambda \sim 1$! In fact, if we set $\lambda^6 = (r^3 - r_-^3)/(r_+^3 - r_-^3)$, identify η with $(1 - (r_-/r_+)^3)^{1/6}$, and take the limit of $r_+ \to r_-$, then (4.10) reduces to (4.8), up to some obvious rescaling. Moreover, according to (4.12), $r_+ \to r_-$ is the limit that T is large, which (as $1/g_4^2 = T/g_5^2$) makes the four-dimensional coupling small.

So in hindsight we could discuss four-dimensional gauge theories in the following way, without passing through the CFT/AdS correspondence. In a spacetime $\mathbf{R}^9 \times \mathbf{S}^1$, consider N Type IIA fourbranes wrapped on $\mathbf{R}^4 \times \mathbf{S}^1$. Pick a spin structure on the $\mathbf{S}^1$ that breaks supersymmetry. This system looks at low energies like four-dimensional $U(N)$ gauge theory, with Yang-Mills coupling $g_4^2 = g_5^2/T = g_{st}/T$. Take $N \to \infty$ with $g_{st}N$ fixed. The D-brane system has both open and closed strings. The dominant string diagrams for large N with fixed $g_5^2 N$ and fixed T are the planar diagrams of 't Hooft [1] – diagrams of genus zero with any number of holes. (This fact was exploited recently [31] in analyzing the beta function of certain field theories.) Summing them up is precisely the long-intractable problem of the $1/N$ expansion.

Now, at least if η is large, supergravity effectively describes the sum of planar diagrams in terms of the metric (4.10) which is produced by the D-branes. This is a smooth metric, with no singularity and no D-branes. So we get a description with closed Type IIA strings only. Thus the old prophecy [1] is borne out: nonperturbative effects close up the holes in the Feynman diagrams, giving a confining theory with a mass gap, and with $1/N$ as a coupling constant, at least for large η. To understand large N gauge theories, one would want, from this point of view, to show that there is no singularity as a function of η, except at $\eta = 0$, and to exhibit asymptotic freedom and compute the masses for small η. (This looks like a tall order, given our limited knowledge of worldsheet field theory with Ramond-Ramond fields in the Lagrangian.) The singularity at $\eta = 0$ is simply the singularity of the fourbrane metric at $r_+ = r_-$; it reflects the classical $U(N)$ gauge symmetry of N parallel fourbranes, which disappears quantum mechanically when $\eta \neq 0$ and the singularity is smoothed out.

I have benefited from comments by N. Seiberg. This work was supported in part by NSF Grant PHY-9513835.

References

[1] G. 't Hooft, "A Planar Diagram Theory For Strong Interactions," Nucl. Phys. **B72** (1974) 461.

[2] A. M. Polyakov, "String Theory And Quark Confinement," hep-th/9711002.

[3] S. S. Gubser, I. R. Klebanov, and A. W. Peet, "Entropy And Temperature Of Black 3-Branes," Phys. Rev. **D54** (1996) 3915.

[4] I. R. Klebanov, "World Volume Approach To Absorption By Nondilatonic Branes," Nucl. Phys. **B496** (1997) 231.

[5] S. S. Gubser, I. R. Klebanov, and A. A. Tseytlin, "String Theory And Classical Absorption By Threebranes," Nucl. Phys. **B499** (1997) 217.

[6] S. S. Gubser and I. R. Klebanov, "Absorption By Branes And Schwinger Terms In The World Volume Theory," Phys. Lett. **B413** (1997) 41.

[7] J. Maldacena and A. Strominger, "Semiclassical Decay Of Near Extremal Fivebranes," hep-th/9710014.

[8] G. Gibbons and P. Townsend, "Vacuum Interpolation In Supergravity Via Super p-Branes," Phys. Rev. Lett. **71** (1993) 5223.

[9] M. J. Duff, G. W. Gibbons, and P. K. Townsend, "Macroscopic Superstrings as Interpolating Solitons," Phys. Lett. **B332** (1994) 321.

[10] S. Ferrara, G. W. Gibbons, and R. Kallosh, "Black Holes And Critical Points In Moduli Space," Nucl. Phys. **B500** (1997) 75, hep-th/9702103; A. Chamseddine, S. Ferrara, G. W. Gibbons, and R. Kallosh, "Enhancement Of Supersymmetry Near $5-D$ Black Hole Horizon," Phys. Rev. **D55** (1997) 3647, hep-th/9610155.

[11] J. Maldacena, "The Large N Limit Of Superconformal Field Theories And Supergravity," hep-th/9711200.

[12] M. J. Duff, B. E. W. Nilsson, and C. N. Pope, "Kaluza-Klein Supergravity," Physics Reports **130** (1986) 1.

[13] M. J. Duff, H. Lu, and C. N. Pope, "$AdS_5 \times \mathbf{S}^5$ Untwisted," hep-th/9803061.

[14] M. Flato and C. Fronsdal, "Quantum Field Theory Of Singletons: The Rac," J. Math. Phys. **22** (1981) 1278; C. Fronsdal, "The Dirac Supermultiplet," Phys. Rev. **D23** (19888) 1982; E. Angelopoulos, M. Flato, C. Fronsdal, and D. Sternheimer, "Massless Particles, Conformal Group, and De Sitter Universe," Phys. Rev. **D23** (1981) 1278.

[15] E. Bergshoeff, M. J. Duff, E. Sezgin, and C. N. Pope, "Supersymmetric Supermembrane Vacua And Singletons," Phys. Lett. **B199** (1988) 69; M. P. Blencowe and M. J. Duff, "Supersingletons," Phys. Lett. **B203** (1988) 203.

[16] S. Hyun, "U-Duality Between Three And Higher Dimensional Black Holes," hep-th/9704005; S. Hyun, Y. Kiem, and H. Shin, "Infinite Lorentz Boost Along The M Theory Circle And Nonasymptotically Flat Solutions In Supergravities," hep-th/9712021.

[17] K. Sfetsos and K. Skenderis, "Microscopic Derivation Of The Bekenstein-Hawking Formula For Nonextremal Black Holes," hep-th/9711138, H. J. Boonstra, B. Peeters, and K. Skenderis, "Branes And Anti-de Sitter Space-times," hep-th/9801206.

[18] P. Claus, R. Kallosh, and A. van Proeyen, "M Five-brane And Superconformal $(0,2)$ Tensor Multiplet In Six-Dimensions," hep-th/9711161; P. Claus, R. Kallosh, J. Kumar, P. Townsend, and A. van Proeyen, "Conformal Field Theory Of $M2$, $D3$, $M5$, and $D1$-Branes + $D5$-Branes," hep-th/9801206.

[19] R. Kallosh, J. Kumar, and A. Rajaraman, "Special Conformal Symmetry Of World-volume Actions," hep-th/9712073.

[20] S. Ferrara and C. Fronsdal, "Conformal Maxwell Theory As A Singleton Field Theory On $AdS(5)$, IIB Three-branes and Duality," hep-th/9712239.

[21] S. Hyun, "The Background Geometry Of DLCQ Supergravity," hep-th/9802026.

[22] M. Gunaydin and D. Minic, "Singletons, Doubletons, and M Theory," hep-th/9802047.

[23] S. Ferrara and C. Fronsdal, "Gauge Fields As Composite Boundary Excitations," hep-th/9802126.

[24] G. T. Horowitz and H. Ooguri, "Spectrum Of Large N Gauge Theory From Supergravity," hep-th/9802116.

[25] N. Itzhaki, J. M. Maldacena, J. Sonnenschein, and S. Yankielowicz, "Supergravity And The Large N Limite Of Theories With Sixteen Supercharges," hep-th/9802042.

[26] S. Kachru and E. Silverstein, "4d Conformal Field Theories And Strings On Orbifolds," hep-th/9802183.

[27] M. Berkooz, "A Supergravity Dual Of A $(1,0)$ Field Theory In Six Dimensions," hep-th/9802195.

[28] V. Balasumramanian and F. Larsen, "Near Horizon Geometry And Black Holes In Four Dimensions," hep-th/9802198.

[29] S. S. Gubser, I. R. Klebanov, and A. M. Polyakov, "Gauge Theory Correlators From Noncritical String Theory," hep-th/9802109.

[30] M. Flato and C. Fronsdal, "Interacting Singletons," hep-th/9803013.

[31] A. Lawrence, N. Nekrasov, and C. Vafa, "On Conformal Theories In Four Dimensions," hep-th/9803015.

[32] M. Bershadsky, Z. Kakushadze, and C. Vafa, "String Expansion As Large N Expansion Of Gauge Theories," hep-th/9803076.

[33] S. S. Gubser, A. Hashimoto, I. R. Klebanov, and M. Krasnitz, "Scalar Absorption and the Breaking of the World Volume Conformal Invariance," hep-th/9803023.

[34] I. Ya. Aref'eva and I. V. Volovich, "On Large N Conformal THeories, Field Theories On Anti-de Sitter Space, and Singletons," hep-th/9803028.

[35] L. Castellani, A. Ceresole, R. D'Auria, S. Ferrara, P. Fré, and M. Trigiante, "G/H M-Branes And AdS_{p+2} Geometries," hep-th/9803039.

[36] S. Ferrara, C. Fronsdal, and A. Zaffaroni, "On $N = 8$ Supergravity On AdS_5 And $N = 4$ Superconformal Yang-Mills Theory," hep-th/9802203.

[37] O. Aharony, Y. Oz, and Z. yin, hep-th/9803051.

[38] S. Minwalla, "Particles on $AdS_{4/7}$ And Primary Operators On $M_{2/5}$ Brane Worldvolumes," hep-th/9803053.

[39] J. Maldacena, "Wilson Loops In Large N Field Theories," hep-th/9803002

[40] S.-J. Rey and J. Yee, "Macroscopic Strings As Heavy Quarks In Large N Gauge Theory And Anti de Sitter Supergravity," hep-th/9803001.

[41] E. Halyo, "Supergravity On $AdS_{4/7} \times \mathbf{S}^{7/4}$ and M Branes," hep-th/9803077.

[42] R. G. Leigh and M. Rozali, "The Large N Limit Of The $(2, 0)$ Superconformal Field Theory," hep-th/9803068.

[43] M. Bershadsky, Z. Kakushadze, and C. Vafa, "String Expansion As Large N Expansion Of Gauge Theories," hep-th/9803076.

[44] A. Rajaraman, "Two-Form Fields And The Gauge Theory Description Of Black Holes," hep-th/9803082.

[45] G. T. Horowitz and S. F. Ross, "Possible Resolution Of Black Hole Singularities From Large N Gauge Theory," hep-th/9803085.

[46] J. Gomis, "Anti de Sitter Geometry And Strongly Coupled Gauge Theories," hep-th/9803119.

[47] S. Ferrara, A. Kehagias, H. Partouche, and A. Zaffaroni, "Membranes And Fivebranes With Lower Supersymmetry And Their AdS Supergravity Duals," hep-th/9803109.

[48] J. A. Minahan, "Quark-Monopole Potentials In Large N Super Yang-Mills," hep-th/9803111.

[49] E. Witten, "Anti-de Sitter Space And Holography," hep-th/9802150.

[50] A. M. Polyakov, "Thermal Properties Of Gauge Fields And Quark Liberation," Phys. Lett. **72B** (1978) 477.

[51] L. Susskind, "Lattice Models Of Quark Confinement At High Temperature," Phys. Rev. **D20** (1979) 2610.

[52] D. J. Gross, R. D. Pisarski, and L. G. Yaffe, "QCD And Instantons At Finite Temperature," Rev. Mod. Phys. **53** (1981) 43.

[53] G. Horowitz and J. Polchinski, "Corresondence Principle For Black Holes And Strings," Phys. Rev. **D55** (1997) 6189.

[54] T. Banks, W. Fischler, I. R. Klebanov, and L. Susskind, "Schwarzschild Black Holes From Matrix Theory," Phys. Rev. Lett. **80** (1998) 226, hep-th/9709091; I. R. Klebanov and L. Susskind, "Schwarzschild Black Holes In Various Dimensions From Matrix Theory," Phys. Lett. **B416** (1998) 62, hep-th/9719108, "Schwarzschild Black Holes In Matrix Theory 2," hep-th/9711005.

[55] A. Sen, "Extremal Black Holes And Elementary String States," Mod. Phys. Lett. **A10** (1995), hep-th/9504147.

[56] A. Strominger and C. Vafa, "Microscopic Origin Of The Bekenstein-Hawking Entropy," Phys. Lett. **B379** (1996) 99, hep-th/9601029.

[57] J. Bekenstein, "Black Holes And Entropy," Phys. Rev. **D7** (1973) 2333.

[58] S. W. Hawking, "Particle Creation By Black Holes," Commun. Math. Phys. **43** (1975) 199.

[59] G. 't Hooft, "Dimensional Reduction In Quantum Gravity," in *Salamfest 93*, p. 284,gr-qc/9310026.

[60] C. Thorn, "Reformulating String Theory With The $1/N$ Expansion," lecture at First A. D. Sakharov Conference on Physics, hep-th/9405069.

[61] L. Susskind, "The World As A Hologram," J. Math. Phys. **36** (1995) 6377.

[62] D. J. Gross and E. Witten, "Possible Third Order Phase Transition In The Large N Lattice Gauge Theory," Phys. Rev. **D21** (1980) 446.

[63] C. Thorn, "Infinite $N(C)$ QCD At Finite Temperature: Is There An Ultimate Temperature?" Phys. Lett. **B** (1981) 458.

[64] S. W. Hawking and D. Page, "Thermodynamics Of Black Holes In Anti-de Sitter Space," Commun. Math. Phys. **87** (1983) 577.

[65] J. W. York, "Role Of Conformal Three-Geometry In The Dynamics Of Gravitation," Phys. Rev. Lett. **28** (1972) 1082.

[66] G. W. Gibbons and S. W. Hawking, "Action Integrals And Partition Functions In Quantum Gravity," Phys. Rev. **D15** (1977) 2752.

[67] J. Cardy, "Operator Content Of Two-Dimensional Conformally-Invariant Theories," Nucl. Phys. **B270** (1986) 186.

[68] A. Strominger, "Black Hole Entropy From Near-Norizon Microstates," hep-th/9712251.

[69] J. D. Brown and M. Henneaux, "Central Charges In The Canonical Realization Of Asymptotic Symmetries: An Example From Three-DImensional Gravity," Commun. Math. Phys. **104** (1986) 207.

[70] M. B. Green and M. Gutperle, "Comments On Three-Branes," Phys. Lett. **B377** (1996) 28.

[71] A. A. Tseytlin, "Self-duality of Born-Infeld Action And Dirichlet 3-Brane Of Type IIB Superstring," Nucl. Phys. **B469** (1996) 51.

[72] S. J. Avis, C. Isham, and D. Storey, "Quantum Field Theory In Anti-de Sitter Space-time," Phys. Rev. **D18** (1978) 3565.

[73] P. Breitenlohner and D. Z. Freedman, "Positive Energy In Anti-de Sitter Backgrounds And Gauged Extended Supergravity," Phys. Lett. **115B** (1982) 197, "Stability In Gauged Extended Supergravity," Ann. Phys. **144** (1982) 197.

[74] E. Witten, "Some Comments On String Dynamics," in *Strings '95*, ed. I. Bars et. al. (World Scientific, 1997), hep-th/9507121.

[75] A. Strominger, "Open p-Branes," Phys. Lett. **B383** (1996) 44, hep-th/9512059.

[76] M. Duff, "Strong/Weak Coupling Duality From The Dual String," Nucl. Phys. **B442** (1995) 47, hep-th/9501030.

[77] E. Witten, "String Theory Dynamics In Various Dimensions," Nucl. Phys. **B443** (1995) 85, hep-th/9503124.

[78] G. Horowitz and A. Strominger, "Black Strings And p-Branes," Nucl. Phys. **B360** (1991) 197.

Can there be Physics without Experiments? Challenges and Pitfalls

Gerard 't Hooft†

Institute for Theoretical Physics
University of Utrecht, Princetonplein 5
3584 CC Utrecht, the Netherlands

and

Spinoza Institute
Postbox 80.195
3508 TD Utrecht, the Netherlands

Summary

Physicists investigating space, time and matter at the Planck scale will probably have to work with much less guidance from experimental input than has ever happened before in the history of Physics. This may imply that we should insist on much higher demands of logical and mathematical rigour than before. Working with long chains of arguments linking theories to experiment, we must be able to rely on logical precision when and where experimental checks cannot be provided.

Introduction.

During the last few decades our view upon matter, the fundamental forces and the geometry of space-time have undergone a metamorphosis, enabling us to extrapolate our knowledge towards time- and distance scales that in fact cannot be directly probed by experiment. This extrapolation leads us towards the so-called Planck scale, where lengths are of the order of 10^{-33} cm, time intervals of the order of 10^{-44} sec, and masses of the order of 20 μg. Here, gravitational effects cause curvature in space and time, and today's physical models, the "superstring theories" and related ideas, tend to become extremely complex. In general, even our book-keeping procedures, needed to formulate the physical degrees of freedom, their coordinates in space and time, and the disciplines for their evolution, present fundamental difficulties.

---

† e-mail: g.thooft@phys.uu.nl; internet: http://www.phys.uu.nl/~thooft/

Physicists will not easily accept any new theory unless it has been extensively tested experimentally and all conceivable alternatives have been ruled out. The extremely successful Standard Model is a perfect example. Being the first complete synthesis of the theories of Special Relativity and Quantum Mechanics, it came into being by a close collaboration between theoreticians and experimentalists.

But with String Theory [1], the situation will be vastly different. Experimentally, this theory will be extremely difficult to assess. Most of our evidence will be extremely indirect at best: the theory shows a slight preference for predicting the emergence of super symmetry at the TeV scale, although this prediction actually was made long before String Theory took its present form. Some cosmological models inspired by String Theory will be experimentally testable but here also the input from these theories is indirect. The most direct predictions refer to the nature of space, time, and the particle spectrum at the Planck scale, where no experiment will ever be able to check any of their claims.

Yet, String Theory is of extreme importance. It is considered by many to be as yet the only detailed avenue towards a better understanding of the synthesis of general Relativity with Quantum Mechanics. One day, we hope to be able to understand the spectral and numerical details of the Standard Model itself. Presently, there are about 26 fundamental constants of nature, whose numerical values we are unable to derive from present theories. It is hoped that String Theory or any other similar construction will do this for us. The question to be discussed now is: Will there be any chance to succeed, if experimental support will be as indirect as we all expect?

To address this question, let us look at the development of important theories in the past, and estimate in what way they owe their existence to the numerous and essential experimental checks.

Theory and Experiment in Physics.

Physics is an experimental science. One studies phenomena observed in our world, ranging from the mundane ones to the extremely esoteric. Obvious, but nevertheless of essential importance are observations such as the fact that our space is three dimensional, and that there exists such a force as the gravitational one, pulling us towards the ground; we see that our universe has matter in it, divided into little particles. It seems that the properties of matter can be reduced to the properties of these particles, which appear to be point like, and their properties, such as the forces acting between them, are universal. Our entire universe appears to be built out of the same material.

The most esoteric observations, requiring the most sophisticated techniques to be realized at all, are for instance the fact that our universe has curvature in it, and that some particles long thought to be massless, actually turn out to have mass after all [2]. Much of the progress in modern physics results from such eminent achievements of the scientific method.

It is the theoretician's task to try to find order in these phenomena. We have experienced that the phenomena obey rules, which we call "laws of physics". It appears that the laws are obeyed with extreme precision, and, most importantly, it seems that some rules are obeyed everywhere, under the most extreme circumstances. They appear to be universal.

An essential complication however is that, at atomic distance and time scales, a new kind of 'fuzziness' arises, allowing us only to apply certain rules of statistics, rather than deterministic rules. This is what we call Quantum Mechanics. But apart from this important side remark, law and order are everywhere.

I am not the first to be surprised by humanity's continuing progress in unraveling these laws, in spite of the fact that these laws do not seem to have been made by humans. Indeed, they appear to be far superior to the man-made laws. History nevertheless demonstrated that we can get into grips with them.

How does this process work? Philosophers of science sometimes give answers of the following sort:

(1) At a given stage of our understanding, there may be competing theories for the explanation of an observed phenomenon, theory #1 and theory #2, say.

(2) Theory #1 leads to prediction #1, and theory #2 leads to prediction #2.

(3) And now an experiment is done. If it agrees with one of the two predictions this helps us to select the right theory, until new phenomena are observed that require further refinements. Go back to (1).

According to this picture, the role of theoreticians is to come with a number of alternative scenarios. They are tested by the experiments. Experiment has the last word. But this picture, often taught at high schools, is not quite the way physicists themselves experience what is really happening. It is far too simplistic to suggest that theoreticians are just sitting there producing theories, preparing them just such that they can be confronted with experiment. The point is that in most cases, the different theories are not "equivalent" at all. They are not equally probable or equally credible. Very often, the choice is between

1) a theory, and 2) no theory.

Or else, the theories are of very different quality. I am talking now of highly 'fundamental' branches of physics, where the laws at the very end of the scales are searched for: either at extremely tiny distance, at the highest conceivable energies, or, conversely, at the largest distance scale. Nowadays we have a very detailed theoretical picture of what is going on, but there are uncertainties, in regions of physics that have been inaccessible in the past.

We try to produce a theoretical scheme that is more advanced, more refined or more ambitious than the more established versions. To find out whether the new version makes sense or not, one naturally first asks whether experiments or new astronomical observations can be envisioned that can give us further guidance. If this is not the case, at least in the foreseeable future, should we then abandon such theoretical research, or do we have guidelines other than experimental verifications?

It turns out that we do. New theories for space, time, matter and forces are being proposed, which, in practice, appear to be such that they may be subjected to tests of the following kinds:

(i) A theory should be *logically coherent.* This demand is far from trivial or obvious. Often, theories are formulated in terms of abstract mathematical equations that ought to control infinitely many degrees of freedom. If these equations display a clear causal structure, and if wild oscillations can be seen to be subject to constraints derived from a bounded quantity such as energy, one might conclude that the theory makes sense logically. However, many complications can arise that imperil logical consistency. If point like particles are replaced by line like objects, locality of the equations is no longer obvious, so that the causal structure may get lost*; if gravitational effects remove the lower bound to the energy, then energy can no longer be used to show the absence of run-away solutions, and if the theory is only formulated in perturbative terms, the existence and internal consistency of a corresponding non-perturbative system may be questioned. And so on.

(ii) The theory should *be capable of* making predictions. Even if the predictions cannot be tested

* In the *perturbative* formulation of String Theory, causality appears to be obeyed perfectly, including its coupling to gravity. It is in the circular nature of any attempt at a non perturbative formulation where reference to time ordering is often lost, so that causality may be questionable.

experimentally, the mere fact that meaningful statements are made concerning the outcome of some *Gedanken*experiment could distinguish a theory from less viable competitors. In short: a theory should at least be capable of predicting coherently the outcome of experiments done in our imagination. Requiring such outcomes to be informative and coherent is an important criterion for our esoterical theory.

(iii) The theory should agree with older theories that are well-established. Thus, most advanced particle theories such as String Theory, M-theory and the like are demanded to agree at least with Quantum Mechanics, and Special and General Relativity. They should allow for the existence of fermions, and a four-dimensional spacetime evolving into something resembling our present universe.

(iv) There should be *agreement* with already observed phenomena. This means that, even if no *new* experiments can be thought of to test our theory, we can still make use of experiments done long ago. For instance, the fact that a gravitatonal force exists, the fact that there are three generations of quarks and leptons, that the effective gauge forces at low energy exhibit the mathematical structure of $SU(3)\otimes SU(2)\otimes U(1)$, all these should be consistent with our theory, and in practice these are very important constraints. Some string theorists proclaim that their theory "predicts" gravity. This is an absurd statement, of course, but it indicates how stringent this requirement appears to be in the eyes of these physicists.

(v) Eventually, a successful theory must be able to make *real* predictions. For string theory, or other considerations concerning physics at the Planck scale, this may mean that we hope to be able, one day, to 'predict' the value of fundamental constants of nature that as yet have eluded any genuine prediction, such as the fine structure constant. Even if this constant could be reproduced with less accuracy than its presently known experimental value, this would be a tremendous achievement. It would imply that, at least in principle, such numbers could be computed with arbitrary precision, so that definitely predictions of the conventional type can be expected. This is surely what we are aiming at eventually. But long before this situation is arrived at, a theory may have become acceptable and respected by a fairly large community, simply because it may have provided more and deeper understanding.

Examples from the past.

It is difficult to compare this situation with our experiences in the past, because physicists always have been so fortunate that clever experiments could be devised to check *directly* whether the equations described in a theory worked the way they were expected to.

Consider the Maxwell equations. James Clark Maxwell discovered around 1864 that the descriptions existing in his time were inconsistent. His equations for electricity and magnetism were a mathematical completion of existing knowledge. He quickly realized that light propagation had to be closely related to electromagnetism, and as the theory agreed with everything known before, it was soon accepted as obeying the first few of the criteria written down above.

Needless to say however, that the theory was tested at numerous occasions when electromagnetic phenomena were further studied. Would we have accepted the theory if such tests had not been there? Clearly, the theory would not have enjoyed the same status as it does today, but the mere fact that, at one stroke, the theory of optics was unified together with electricity and magnetism would have made it highly respectable.

Einstein's theory of relativity was just such a case. The theory was not primarily intended to "explain" observed phenomena. Although Einstein presumably knew about the Michelson-Morley experiments, his primary concern was to give a more satisfactory description of the geometry of space and time in its relation to electro-magnetism. His analysis was a purely mathematical one. His successful attempts to include the gravitational force, led to the generalized theory of relativity, and this happened to provide for an unexpected explanation of an observation that had already been made but was not understood until then: the anomalous shift of the orbit of the planet Mercury. It was also very fortunate that Einstein's theory gave rise to an opportunity for further experimental tests, notably the famous Eddington expeditions.

In this case, the role of the experiment was not to discriminate between Einstein's theory and some would-be alternative; there hardly were alternatives‡, but the theory could have been wrong or incomplete for other reasons. The test was made in order to find new confirmations of the fact that Physics here was on the right track. this is a situation that we see occur quite often in our field of science. Historically, these tests were generally thought to be of extreme importance. What this means, is that theories that are so novel as relativity was at that time, cannot flourish without as much support as possible from independent sources of information.

Many of our present theories are to be compared to relativity *without* such checks. The builders of these theories are facing hard times; yet I believe that we *can* make progress, provided our techniques, as well as our critical attitude regarding the required logical consistence, improve.

A milestone in physics was Paul Adrien Maurice Dirac's improved equation for the electron. Previously, Wolfgang Pauli had the following (non-relativistic) equation:

$$H\psi = \Big(\frac{1}{2m}(\hat{\mathbf{p}} - e\mathbf{A})^2 - \frac{g\,e}{2m}(\mathbf{B}\cdot\boldsymbol{\sigma})\Big)\psi\ ,$$

where the last term describes the interaction between the electron's magnetic moment with the magnetic field $\mathbf{B}$. It features an unknown constant, the electron's gyro-magnetic ratio, g. Now Dirac's equation [4] was

$$H\psi = \Big(\boldsymbol{\alpha}\cdot(\hat{\mathbf{p}} - e\mathbf{A}) + \beta m\Big)\psi\ ,$$

where $\boldsymbol{\alpha}$ and β are the Dirac matrices.

In the non-relativistic limit, Dirac's equation turns into Pauli's equation, but it offers a bonus: the gyromagnetic ratio g is predicted to be exactly 2, which was very close to the experimentally observed value. However, Dirac realized very well that there also was a difficulty with his equation. His electron is described by 4 components, not 2, as in Pauli's equation. The two extra degrees of freedom could be interpreted as describing fermions with electric charge $+e$ instead of $-e$, as in the electron. About this episode, A. Pais writes [5]:

> While in Leipzig, June 1928, Dirac visited his friend and colleague Werner Heisenberg, who had recently been appointed there. Heisenberg was also well aware of the difficulties. In May he had written to Pauli: "In order not to be forever irritated with Dirac, I have done something else for a change" (the something else was his quantum theory of ferromagnetism). Dirac and Heisenberg discussed several aspects of the new theory. Shortly thereafter, Heisenberg wrote again to Pauli: "The saddest chapter of modern physics is and remains the Dirac theory." He mentioned some of his own work which demonstrated the difficulties, and added that the magnetic electron had made Jordan "trübsinnig" (melancholic). At about the same time, Dirac, not feeling so good either,

‡ There *do* exist various kinds of *modifications* of Einstein's theory, such as Dicke's scalar-tensor theory, which could be tested by these experiments — they were all but ruled out — but even these were based on Einstein's principle of invariance under general coordinate transformations. There actually also existed an alternative explanation for Mercury's perihelion shift: an unknown planet very near to the sun...

> wrote to Oskar Klein: "I have not met with any success in my attempts to solve the $\pm e$ difficulty. Heisenberg (whom I met in leipzig) thinks the problem will not be solved until one has a theory of the proton and the electron together."

Indeed, during some time, Dirac and his coleagues suspected that the new components had to be describing the proton, but this gave rise to even more difficulties: Oppenheimer [6] and Tamm [7] correctly objected that, if this were true, it would make the hydrogen atom unstable against decay into two or more photons. Finally, in 1931, Dirac had to admit, much against his will, that his equation predicted a particle not observed until then. Just before the year's end, Carl Anderson made the first announcement of experimental evidence for the anti-electron [8].

What this history tells us is that the relation between theory and experiment is quite a lot more subtle than a game of differentiation and rejection of false theories. Theories are more often than not rejected solely on the grounds that they do not seem to be reconcilable with known facts.

Another case of interest is Pauli's theory of the existence of a "neutrino" to account for the missing energy in β decay. In February 1929, Pauli complained in a letter to Oskar Klein that Bohr was serving him up all kinds of ideas about beta decay

> "...by appealing to the Cambridge authorities but without reference to the literature", and he said: "With his considerations about a violation of the energy law Bohr is on a *completely wrong* track" [5].

The existence of neutrinos was confirmed by Frederick Reines' experiment [9] at the Savannah River reactor, in 1956. When he Announced the discovery in a telegram to Pauli, the latter reacted not the least surprised: "I knew already that they exist".

Again a different situation occurred when the new theories for the weak interaction were proposed. In the late '60s the problem was that the existing theories for weak interactions were simply too diffuse to be believable.

When C.N. Yang and R.L. Mills did their work on the generalization of QED equations, in 1954, the paper they published [10] contained a theory that *did not seem to apply to any known situation* in particle physics. Yet, the theory had such a high degree of elegance that they believed it to contain some deep truth. Although this new theory was ignored by many physicists, it did provide inspiration to some of the deeper thinkers. Richard Feynman used it at various occasions: the Feynman-Gell-Mann expression [11] for an effective weak interaction theory (a refinement of an earlier proposal by Enrico Fermi), carried some elements of the Yang-Mills idea, and somewhat later when Feynman made an attempt to understand the quantization of the gravitational force, he was made aware by Gell-Mann of the deep mathematical relations between gravity and Yang-Mills [12]. So a theory without any experimental verification nevertheless played a role. Insignificant as it seemed to be at the time, this role would later turn out to be an essential one.

In 1961, C.N. Yang was asked to lecture about "The Future of Physics" [13.], at a panel discussion at the Celebration of the MIT Centennial, April 8. At that time, this is what Yang dared to say about his expectations:

> "[What I know now, does] enable us to say with some certainty that great clarification will come in the field of the weak interactions in the next few years. With luck on our side, we might even hope to see some integration of the various manifestations of the weak interactions."

Yang hoped, *with luck on his side*, to bring all *weak* interactions under one denominator. The fact that, within ten years, one single theory, based on his own ideas, would not only achieve that, but also include electromagnetism, apparently was way beyond his hopes.

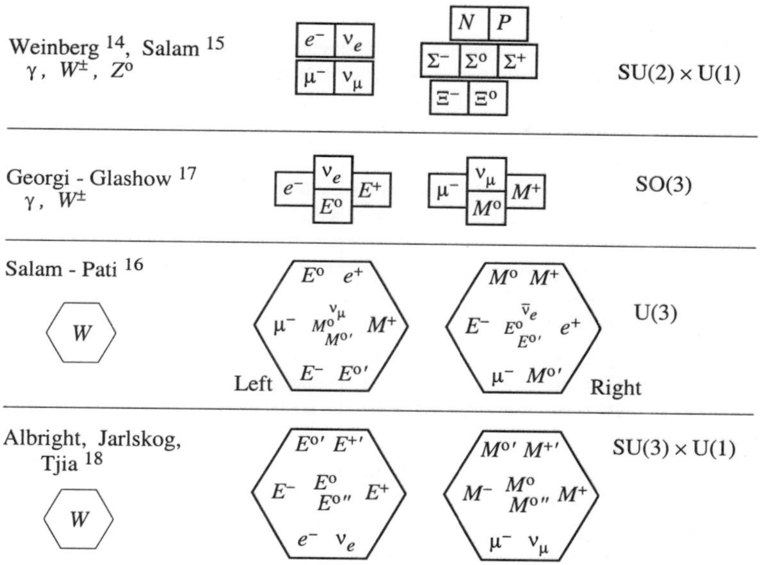

Fig. 1. A few weak interaction models of the early '70's, showing some of the proposed multiplet structures. E and M are proposed heavy leptons. Also indicated are the predicted vector bosons. The hexagons labe6led "W" indicate nonets of charged and neutral vector bosons. For the last three models only the leptonic sectors are displayed.

In the early 70's it was discovered that *theories of this type were indeed logically fully consistent*, and superior to any of their alternatives. But it was also discovered that a whole class of such theories could be constructed, all having the Yang-Mills equations as a back bone. The earliest, already existing model was due to Steven Weinberg [14]. and Abdus Salam [15], and it was based on $SU(2) \otimes U(1)$. But among the alternatives was a model proposed by Pati and Salam [16], whose version was based on $SU(3)$, and a more interesting alternative theory was proposed by H. Georgi and S.L. Glashow [17], based on the group $SO(3)$. See Fig. 1, where some of the models that were proposed are illustrated.

Most of the models predicted not only the charged vector bosons that intermediate the observed weak interactions, but also one or more neutral components, called Z^0. Such particles would cause interactions between neutral components of the currents in hadrons and leptons. Such interactions had not yet been detected. What made the Georgi-Glashow model interesting was that, having no Z^0 boson, it would predict the *absence* of any such effects. In the '70's, a combined effort by several experimental teams (CERN [19], Fermilab [20], Argonne [21], Brookhaven [22]) reinvestigated the spurious weak interaction events associated to the neutral current interactions. Notably, according to most of the new theories, a neutrino should be able to hit against hadrons as well as leptons without itself changing into a charged lepton. Actually, such events had been searched for earlier, with apparently negative results: common wisdom had been that these events do not occur. But it was also agreed upon that these investigations were not totally conclusive, as the events searched for could easily have been mistaken for background noise.

Around 1974 the results of these experimentations became clear: the events were there, more or less as they were predicted by the simplest of all models: the Weinberg-Salam scheme. For theoreticians and experimentalists alike, this was decisive for selecting the original Weinberg-Salam scheme as the most likely theory for the electro-weak interactions. Could we have *derived* from purely theoretical arguments that the model was superior?

The Weinberg-Salam scheme did require a special mechanism, called the GIM mechanism [23], to cancel out all those weak interaction effects that are associated to a change of strangeness, without the exchange of electric charge ('strangeness changing neutral current events'). It was well-known that such interactions, if they existed at all, were very much weaker than the other weak interactions. Since this mechanism required the existence of a new quark species, the charmed quark, the search for this object began, leading to its discovery about a year later. What would have happened if we had not had all these experimental clues?

The alternative models did not describe the hadronic sector very well. In particular, they would have been very difficult to apply to the quark constituents of hadrons. Only for the Weinberg-Salam model did we have a scheme, the GIM mechanism. This had given the model an edge, even before the results of the experiments were known.

One crucial prediction was essentially missed: the new hadronic objects that can consist of only charmed quarks would turn out to have quite conspicuous properties. Experimenters justifiably sometimes ignore theory, by just searching for interesting things even if not explicitly predicted. Some theoreticians did indeed urge to search for narrow resonances, but these were not really considered to be so pronounced or special that they would have striking properties, that would stand out so much that a new field of physics would emerge. But we could have known. When these objects were discovered, called J (by S. Ting [24]) and ψ (by B. Richter [25]), in 1974, they were striking indeed. Most remarkable was the fact that their strong interaction properties were far weaker than normal. It was only during a short time that several "alternative theories" were considered to explain J/ψ. For instance, it was suggested that they might be the bosons intermediating the weak interactions, or maybe they were totally new objects of physics. Several alternative theories were concocted, completely in Popperian style. Yet the charm theory stood out from the beginning. After half a year it was the only theory.

Since that time, theoretical physics has enormously gained in importance. What we have presently is an extremely detailed description of all particles and forces known to us. It is called the "Standard Model". There is actually only a limited amount of information that is still waiting to be provided by experiments in the near future: more details about the neutrino masses and their mixings [2], more precise determination of the mixing angles between quarks, responsible for the violation of PC invariance and top and bottom decays, and the mass (and possibly the spectrum) of the Higgs particle.

Of course we hope that accelerators in the near future will allow us to find further extensions of the Standard Model. Indeed, extensions are expected: supersymmetric versions of the theory (the Minimally Supersymmetric Standard Model, or MSSM) require the existence of much more elementary particle types than the ones known presently: the "super partners". Again, this is the presently preferred scenario, to be confirmed, or to be falsified, in which latter case we have not much to put in its place, other than "new and unexpected physics". Since I do have certain objections against the super symmetric scenario, I would love to see a more complicated physical scene than the one predicted by the MSSM, but detailed predictions cannot be provided.

The reason for giving this talk, and for emphasizing the important shifts in the relations between theory and experiment, is my concern over the situation my field of research will be in, after these facts may or may not have been checked by experiments. We are rapidly proceeding with our theories. Our theoretical view of the particle scenery is presently so clear that we can project much further than the supersymmetry domain. Most importantly, we want to understand the gravitational force.

It is here that possibilities for experiment are severely limited. The Planck scale is determined by the combinations of the constants c (the velocity of light), $\hbar$ (Planck's constant divided by 2π), and G (Newton's constant) that are listed in Table 1.

$$\sqrt{\frac{G\hbar}{c^3}} = 1.616 \times 10^{-33} \text{ cm}$$

$$\sqrt{\frac{G\hbar}{c^5}} = 5.39 \times 10^{-44} \text{ sec}$$

$$\sqrt{\frac{\hbar c}{G}} = 21.77\ \mu\text{g} \rightarrow 1.22 \times 10^{28}\ e\text{V} .$$

Table 1. The Planck Units.

Several theoretical approaches are addressing the fundamental questions we are confronted with here, but of these, Superstring Theory, and its successors (D-brane theory and M-theory), are taking the lead. String Theory started as an attempt to understand the strong interactions, but its mathematical elegance became as striking as that of Yang-Mills theory, 20 years earlier. Here again, theoreticians were dealing with a general scheme that inspires us towards finding deeper connections between space, time and matter. The theories are gaining in power now that various extremely essential difficulties are being confronted and new resolutions are being discovered. The theoretical scheme now being devised is generally believed to be the "only credible theory that unifies the gravitational forces with other forms of matter in a quantum theory".

Most important here is the fact that the gravitational force is a *cumulative force*, which means that, as it acts, it may draw large amounts of matter closer together, and in so doing create regions of space-time where the field becomes increasingly stronger. This makes gravity itself *fundamentally unstable*: it may give rise to "gravitational collapse", or in more popular terms: a black hole.

All theoretical scenarios agree that, at least in principle, initial state conditions can be realized that inevitably lead to gravitational collapse: black holes, or objects very closely related to them, must be part of the matter spectrum at the Planck scale.

The black hole is a prototype of a physical system that can only be theorized about, but not be studied experimentally; nobody ever saw a black hole up close. Here in particular, one must ask the question whether the existing theories describing their behaviour are fully self-consistent, and whether these descriptions give rise to any kind of logical conflict. Indeed, attempts to arrive at a consistent description using well-known and accepted ingredients of physics, have led us to quite peculiar results.

A black hole appears to behave as if it is a "ball of material" that carries one "bit of information" in every $0.724 \times 10^{-65}\ \text{cm}^2$ of its surface area. At the same time, however, a region of space-time that undergoes gravitational collapse into a black hole must also be able to describe ordinary particles living there, and, at first sight, it seems as if these particles are not constrained by a limit on the information they carry. This at least is what we appear to arrive at when applying Einstein's theory of general relativity. On the other hand, black holes appear to drain away information from space-time, whereas in general such behaviour would be in conflict with standard formulations of quantum mechanics [26].

How can one incorporate these apparently conflicting features correctly in a theory? Remarkably, string theoreticians recently succeeded in describing black holes (of certain types) in such a way that their results agree completely with everything just said above, so that their theory appears to be in good shape. [27]

Yet there are still numerous difficulties. Not the least of these is that theories of this nature do not provide simple answers to the question as to how they would predict the observed features of the Standard Model, or how in general any detailed calculation should be done to confront the theory with experiment at low energy scales.

There is no *direct* experiment to test any of the more esoteric ideas. This is not to say that there would not be *indirect* clues. There are still quite a few possibilities for future experiments and observations. The determination of the neutrino mass spectrum is just one of these. Clarification of the supersymmetric domain (if it exists!) and possible detection of superweakly interacting massive particles (if they exist!) are other possibilities. Furthermore, we may hope that astrophysical observations such as the spectrum and fluctuations of the cosmic microwave radiation will yield further data. But, to extrapolate from any of such observations all the way to a string theoretical description at the Planck scale, will continue to be a task of gargantuan proportions.

The only instrument we can use to study the Planck length in detail, is our minds.

In previous years we had experiment to guide us and to protect us against our numerous mistakes, so the correct theories could be identified and confirmed. This, as it turns out, was a luxury that we may have to do without in the future. Does such physics make sense?

Well, we have seen that, more often than not, experiment was not the only way to check a theory. It was logical consistence, general esthetics, agreement with previously known facts, and general rigour, that we could use as criterea in addition.

Unfortunately, history has also shown how easily we can be misled into wrong theories, and how important it was, throughout the ages, that we could ask Nature to settle our disputes. Numerous publications on the new esoteric theories show quite clearly that the moderating control of experiment has been absent for some time. The mathematics is often not as rigorous as one should wish. What disturbes me most is that main progress seems to be coming from theories that take the form of "conjectures". Failing to falsify the conjectures is often presented as a proof of their validity. In reality, these conjectures are essentially impossible to prove (or to disprove) with any kind of rigor since the theories are not sufficiently well defined. How sure are we that our logic is correct? In the past, such a situation has often showed to be detrimental for a theory.

On the other hand, if the logic in an argument appears to be less than perfect, this is not at all a sufficient reason to abandon further pursuit along the lines given. Remember that renormalization theory appeared to have serious shortcomings, yet it worked, just because the difficulties could be kept under control.

Do we have the difficulties of string theory under control? My presently most favored approach is that the most celebrated edifice of Physics, the theory of Quantum Mechanics, must undergo revision. [28] I am not the only one saying this, but I do think that I have a better way to write improved theories and to show why they are superior, and furthermore that I have indications as to *why* Nature chose to run in accordance with laws that feature the remarkable principles of statistics that characterize Quantum Mechanics.

It was a momentous achievement of a century of Physics to find how the dynamics of molecules, atoms and subatomic particles could be described in detail using the principles of Quantum Mechanics. It is conceivable that the next century will tell us how to reconcile these findings with a deterministic world view. All that is needed is a description of a ***primordial basis*** of states,

$$|\psi_1\rangle,\ |\psi_2\rangle,\ \dots$$

in terms of which the evolution operators U behave as pure permutation operators, not just any set of unitary operators. A primordial basis may be seen to be generated by operators $\mathcal{O}(\mathbf{x}, t)$ which commute at all times:

$$[\mathcal{O}(\mathbf{x},\, t),\ \mathcal{O}(\mathbf{x}',\, t')] = 0, \qquad \forall\, \mathbf{x}, \mathbf{x}', t, t'\ ,$$

in contrast with more conventional fields, which do not commute within the light cones. Physicists are presently used to the idea that such operators $\mathcal{O}$ cannot possibly exist, but this is not totally obvious. When gravitational forces are taken into account, any 'no-go' theorem would be extremely difficult to prove.

Naturally, my critical audience asks what, in such a theory, would be the cause and the meaning of interference effects. A possible answer is that interference is only recognized as such *after* having assumed some law of physics describing particles in semi classical terms as objects that can be detected in an experiment. These laws of physics may in reality be extremely simplified versions of a statistical analysis using only a very small subspace of Hilbert space generated by the primordial states: the states with very low energies. This subspace is built from superpositions of the primordial basis elements, being the low-energy eigenstates of the hamiltonian (defined to be the true operator generating time shifts). Since these are *not* primordial states, physicists experience interference effects all the time.

The theory, of which the above is only a very brief outline, leads to several difficulties that are as yet far from being resolved. A primary difficulty is the simple observation that, in the real world, the hamiltonian happens to have a lower bound. If I define the hamiltonian to be the generator of time shifts, then the presence of a lower bound is difficult to reconcile with determinism. A possible way out of this dilemma was recently identified: information loss.

If ideas of this sort could be combined with string theory, this might lead to the insights still needed to produce the necessary logical coherence in our theories. But will it be enough?

Without much guidance from experiment, obtaining better understanding of Nature may perhaps be prohibitively difficult. But my confidence in human ingenuity is tremendous. Even if experiments will be difficult, smart new ideas for experiments will still be proposed, but they will require the utmost skill for a correct interpretation. To do our job, to reach the ultimate equations of physics, with the given limitations for experimental evidence, will probably take much more time than the 20 years that were S.W. Hawking's estimate [29]. During this time, probably many new generations of young physicists will enjoy the fine taste of new discoveries. We are all full of optimism for the new century.

References

1. M.B. Green, J.H. Schwarz and E. Witten, *Superstring Theory*, Cambridge Univ. Press; J. Polchinski, *String Theory*, Cambridge Univ. Press, 1998

2. Y. Fukuda et al (Super-Kamiokande Collaboration), *Phys. Rev. Lett.* **81** 1562 (1998); H. Terazawa, KEK Prepr. 98-226, *Proc. Int. Conf. on Modern Developments in Elementary Particle Physics*, Cairo Helwan and Assyut, Jan and Feb 1999.

3. A. Pais, *Subtle is the Lord...*, Oxford University Press, Oxford, 1982, pp. 115-119.

4. P.A.M. Dirac, *Proc. Roy. Soc.* **A117** (1928) 610; *ibid.* **A118** (1928) 315.

5. A. Pais, *Inward Bound, Of Matter and Forces in the Physical World*, Clarendon, Oxford University Press, New York, 1986, p. 309, 346, ...

6. R. Oppenheimer, *Phys. Rev.* **35** (1930) 562.

7. I. Tamm, *Zeitschr. f. Phys.* **62** (1930) 545.

8. C.D. Anderson, *Science* **76** (1932) 238.

9. C.L. Cowan et al, *Science* **124** (1956) 103.

10. C.N. Yang and R.L. Mills, *Phys. Rev.* **96** (1954) 191, R. Shaw, Cambridge Ph.D. Thesis, unpublished.

11. R.P. Feynman and M. Gell-Mann, *Phys. Rev.* **109** (1958) 193.

12. R.P. Feynman, *Acta Phys. Polonica* **24** (1963) 697.

13. C.N. Yang, *The Future of Physics*, in *Selected Papers 1945-1980 With Commentary*, C.N. Yang, W.H. Freeman and Co, 1983, ISBN 0-7167-1406-X, p.319.

14. S. Weinberg, *Phys. Rev. Lett.* **19** (1967) 1264; id., *Sci. Am.* **231** (1974) 50.

15. A. Salam and J.C. Ward, *Phys. Lett.* **13** (1964) 168.

16. J.C. Pati and A. Salam, *Phys. Rev.* **D8** (1973) 1240.

17. H. Georgi and S.L. Glashow, *Phys. Rev. Lett.* **28** (1972) 1494.

18. C.H. Albright, C. Jarskog and M.O. Tjia, *Nucl. Phys.* **B86** (1975) 535.

19. C. Rubbia, *Proc. of the XVIIth Int. Conf. on High Energy Physics*, London 1974, p. IV-114, 117; J. Sacton, *ibid.*, p. IV-121 ; A. Rousset, *ibid.* p. 128.

20. B.C. Barish et al, *Proc. of the XVIIth Int. Conf. on High Energy Physics*, London 1974, p. IV-111; *Proc. Int. School of Subnuclear Physics, Erice, 1975*, A. Zichichi, ed., Plenum 1977, p. 897.

21. P. Schreiner, *Proc. of the XVIIth Int. Conf. on High Energy Physics*, London 1974, p. IV-123.

22. W. Lee, *Proc. of the XVIIth Int. Conf. on High Energy Physics*, London 1974, p. IV-127.

23. S.L. Glashow, J. Iliopoulos and L. Maiani, *Phys. Rev.* **D2** (1970) 1285.

24. C.C. Ting, *Proc. Int. School of Subnuclear Physics, Erice, 1975*, A. Zichichi, ed., Plenum 1977, p. 559; J.J. Aubert et al, *Phys. Rev. Lett.* **33** (1974) 1404.

25. J.-E. Augustin et al, *Phys. Rev. Lett.* **33** (1974) 1406.

26. G. 't Hooft, *J. Mod. Phys.* **A11** (1996) 4623; gr-qc/9607022.

27. J. Maldacena, *Black Holes in String Theory*, hep-th/9607235; A.W. Peet, *Class. Quant. Grav.* **15** (1998) 3291.

28. G. 't Hooft, *Quantum Gravity as a dissipative deterministic system*, gr-qc/9903084, *Class. Quant. Grav.*, to be publ.

29. S.W. Hawking, in several of his recent public lectures.